The Origin of Modern Humans

This delicately carved, 25,000-year-old ivory head, known as the Venus of Brassempouy, from Laudes, France, shows great detail and is highly realistic.

THE ORIGIN OF MODERN HUMANS

Roger Lewin

SCIENTIFIC
AMERICAN
LIBRARY

A division of HPHLP
New York

Library of Congress Cataloging-in-Publication Data

Lewin, Roger.
    The origin of modern humans / Roger Lewin.
      p. cm.
    Includes bibliographical references and index.
    ISBN 0-7167-5039-2
    ISBN 0-7167-6023-1 (pbk)
    1. Man—Origin. 2. Human evolution. I. Title.
    GN281.L55 1993   93-17647
    573.2—dc20   CIP

ISSN 1040-3213

ISSN 1040-3213

Printed in the United States of America

Scientific American Library
A division of HPHLP
New York

Distributed by W. H. Freeman and Company
41 Madison Avenue, New York, NY 10010
Houndmills, Basingstoke RG21 6XS, England

To the memory of Allan Wilson

CONTENTS

PREFACE

Fortuitous turns of events have shaped my life, as I suspect is the case for most of us. So it was that almost twenty years ago an assignment from a publisher took me into paleoanthropology. I have been privileged these past two decades to have been a close spectator of developments in the field and have perhaps played a role as a commentator. Furthermore, my original training in biochemistry and evolutionary biology has proved invaluable over the years, as the science of anthropology has expanded to include more input from these areas.

In all sciences, prevailing theories are the result of a historical process of continual testing and reformulation. The accumulation of new evidence plays an important role in this process, of course. So too do changes in ideas about mechanisms—in this case, the mechanisms of population biology and evolution. Such changes may influence the interpretation of existing evidence, so that theories may progress in the absence of new evidence. In anthropology, a third contributing factor to the status of prevailing theory is more sociological and concerns the perception of the subject of study: ourselves.

The Copernican and Darwinian revolutions dislodged humans from a position of centrality in the universe of things. Nevertheless, even if humans are accepted as the product of an evolutionary process in common with other species, it is still possible to view *Homo sapiens* as a special product of that process and indeed as its ultimate goal. Such a view obtained in the early decades of this century and, though not explicitly stated or as strongly felt, may be seen to influence current debate. Perhaps it always will.

With the addition of newly developed scientific techniques to the study of the origin of modern humans—*Homo sapiens*—the issues involved have become more sharply formulated. Nevertheless, a heritage of earlier ideas continues to inform the way researchers approach their science as they tackle this question. For this reason, I have included substantial coverage of the science's history with respect to the origin of modern humans. Fascinating in itself, it is also relevant to understanding the modern debate.

Without doubt, one of the more important aspects of the current discussion of the origin of modern humans is the impact of genetic evidence. Molecular anthropology, as the marriage between molecular biological techniques and anthropology has been termed, is more than two decades old, but in fact it is still in its infancy. The approach, though potentially powerful, is highly complex and open to misinterpretation. As a result, it has become extremely controversial.

Molecular and fossil evidence are very different: one looks from the present into the past, the other from the past toward the present. Ultimately, the answers they provide must be the same. The next two decades will surely move toward a resolution of the conflict between these two approaches to yield a clearer perspective on the pattern of evolution that resulted in *Homo sapiens*. The field is currently divided into two very different views; one of them must be wrong, and one day will be demonstrated to be so. Or, perhaps both will be proved to be partially correct, with a middle-ground solution prevailing. The excitement of the science is both the cut and thrust of ideas, and the ultimate question itself: how did humans come to be the way they are?

Any individual's perspective on a topic as fluid and as marked by deep differences of professional opinion as this one is certain to be singular in some way. I have tried to present all sides of the present discussion, though I do not pretend not to hold an opinion. I hope readers will be able to arrive at their own conclusions.

I should like to acknowledge Jerry Lyons, who encouraged me to venture into this controversial intellectual territory. Jerry, who is surely one of the more perspicacious editors in the world of science publishing, has identified a subject worth writing about. Amy Johnson wielded a firm but friendly guiding hand with the high standards of The Scientific American Library. And Travis Amos and Susan Moran brought the volume to life with a thoughtful program of illustration. Once again I am pleased to recognize my debt to my many anthropological friends, some of whom also helped this endeavor by supplying photographs.

The Origin of Modern Humans

PROLOGUE

Sixty thousand years ago, in the rugged Zagros Mountains of northern Iraq, an old man died. He was no ordinary man, however. He was a medicine man, a shaman, the people's link to worlds beyond the tangible, the spirit worlds. Mourning the loss of a revered member of their community, the people of Shanidar prepared for a special burial.

Young men who, if they master the rituals of their calling and the technique of trance, would one day be shamans themselves went out into the mountainside meadows to collect flowers, some brightly colored, others relatively inconspicuous: yarrow, St. Barnaby's thistle, groundsell, grape hyacinth, hollyhocks, cornflowers, a kind of mallow, and horsetail. It was late spring, and the profusion of montane flowers belied the frigid grip with which the Ice Age still held the Earth—and would for a further 50 millennia. Higher up the slopes, forests of firs stood dark green against the strong blue sky, while the valleys below sheltered copses of ash and alder.

Neanderthal skull (Shanidar I), unearthed by Ralph Solecki in April 1957 from the Shanidar Cave in Northern Iraq. The damage to the skull (left side, not visible) was probably done by postmortem rock fall in the cave.

At the cave, the young men arranged the flowers they had collected, forming a bed with the bundles of woody horsetail branches, the last resting place for the shaman. Festoons of yarrow, hollyhock, and other flowers were woven among the ramose stems of the woody horsetail. It was to be a colorful and symbolic spectacle, for the flowers had been selected not only for their appearance but also for their medicinal qualities, the natural remedies that the shaman had exploited in performing his numinous role in society. Another shaman would carry out the final rituals for the dead man, inviting the ancestral spirits of Shanidar to receive so important a soul as the dead shaman's, chanting phrases that, in cryptic form, told the story of the creation of the Shanidar people, their special place in the world, the inevitability of life and death—a never-ending cycle established by the ancient spirits of Shanidar.

The Reality of Shanidar

Fiction, of course, but based on elements of fact. The body of an old man did come to rest in the Shanidar Cave 60,000 years in the past. His bones were excavated three decades ago during an expedition led by Ralph Solecki of Columbia University. Flowers of the type described were strewn under and around his body. This at least is the supposition of Arlette Leroi-Gourhan, of the Musée de l'Homme, in Paris, who identified pollen from those flowers in soil samples taken from around the skeleton. The local vegetation was as described. Beyond this, all is speculation, an attempt to cover the ancient bones with some semblance of what may be recognized as human. "We are brought suddenly to the realization that the universality of mankind and the love of beauty go beyond the boundaries of our own species," wrote Solecki, describing the "flower burial" in his book *Shanidar.* "No longer can we deny the early men the full range of human feelings and experience."

Top: *Workers using the bucket-brigade method for removing debris from the Shanidar cave during the early 1950s; an automatic conveyor is also operating.* Bottom: *The skeleton of the so-called flower burial (Shanidar IV) was excavated by T. Dale Stewart in 1960. Soil from around the skeleton, that of an old man, contained fossilized pollen.*

The old man of Shanidar was a Neanderthal, stockily built, powerfully muscled, with a characteristic long, low skull and protruding face. He was not like anyone on Earth today—not a modern human. Some anthropologists argue that Neanderthals' biological proximity to modern humans should be recognized by calling them *Homo sapiens neanderthalensis*, a subspecies of *Homo sapiens*. Others insist that the differences should be stressed; they prefer the species name *Homo neanderthalensis*. Whatever the reality of Neanderthals' biological relationship to modern humans, Solecki was convinced that the old man of Shanidar enjoyed "the full range of human feelings and experience." Neanderthals, Solecki seems to be saying, were just like us, and yet were not us: physically they were different, but they were human inside. This ambiguity captures something of the ambivalence that anthropologists have manifested over the nature of Neanderthal Man ever since his fossil bones were first discovered, close to a century and a half ago.

Neanderthals and Others

The job of anthropologists is to explain how *Homo sapiens* came to be, physically and mentally. The annals of the science record the ebb and flow of ideas as theories change and new evidence comes to light. In the midst of it all the Neanderthals have been a constant presence, albeit with fluctuating prominence. Discovered in August 1856, a little more than three years before Charles Darwin's *Origin of Species* was published, Neanderthal Man was the first early human type to be recognized. Clearly similar to modern humans in many anatomical respects, including a brain marginally bigger than that of modern humans, the Neanderthals presented anthropologists with a challenge: Were these fossil people our direct ancestors, or were they a side branch on the human family tree? Even though the eminent British anthropologist

Sir Grafton Elliot Smith said in 1928 that the status of Neanderthal Man had been "definitively settled by the investigations of Schwalbe in 1899," there is still disagreement over the matter.

The prominence Neanderthals enjoy in this episode of human history is understandable, given that of all types of fossil humans that are known, Neanderthal discoveries have been by far the most numerous. Fossil bones from at least 200 individuals have been recovered during the past one hundred and forty years, including a dozen almost complete skeletons—a great rarity in the human prehistoric record. Their anatomical appearance arrests our imagination. They were the quintessential cavemen, powerfully built, wresting a living from the most adverse of environmental circumstances. The evidence that in some cases they buried their dead—as in the putative burial of the old man of Shanidar—is, however, their most poignant link with us across evolutionary time.

The Neanderthals were not, however, the only form of premodern human in the world at the time—that is, between about 150,000 and 34,000 years ago. Nor did the Neanderthals occupy all habitable parts of the Old World: they were restricted principally to western Europe and the Near East. Other premodern humans lived throughout eastern Europe, East Asia, and Africa at this time. Heavily built like Neanderthals, these other premodern people did not display the extremes of anatomy, particularly in the face, that characterize Neanderthals. Collectively, all these populations (including the Neanderthals) have been termed archaic *sapiens*. A very much sparser fossil and archeological record associated with the eastern European, East Asian, and African premodern people means that much less is known about them than has been documented for the Neanderthals. When the question "What was the fate of the Neanderthals?" is asked, in connection with the origin of modern humans, however, the same question must also be asked of these other premodern people. They are all part of the same evolutionary history: that is, the origin of modern humans.

Evanescent Evidence

An understanding of the evolution of any species—man or mouse—involves three elements: pattern, process, and content, or biology. Pattern describes the phylogenetic history of a species—the family tree, in popular terminology. Process is concerned with the mechanisms by which branches originate or terminate on that tree. And biology refers to the lifeways of a species: how it moves about, its nutritional strategy, its social structure, its cognitive abilities—in short, how it behaves in the world. Although pattern is the most direct inference that can be made from the fossil record, the less material aspects of our history are often what most firmly hold our interest in the evolu-

tion of modern humans. Language, consciousness, mythology—these are the elements that make us human, the elements to which Solecki was appealing in his assessment of the Shanidar people.

Solecki did not adduce standards of stone-tool manufacture, for instance, or evidence of sophisticated subsistence strategies to support a claim for humanness in the Neanderthals. Instead, he pointed to flowers at a putative ritual burial. Solecki was lucky in his discovery—if indeed his interpretation is correct—because often those actions that most suggest humanity are least visible in the prehistoric record. Consider, for instance, the Australian Aborigines. Theirs is a rich social and religious tradition, with kinship systems and mythologies as complex as can be encountered in any foraging society. Their symbol-

The evident richness of Australian Aboriginal ritual, expressed through body decoration, painting, and dance, is a reminder of how little of such behavior becomes imprinted on the archeological record.

ism is arrestingly visual too, manifested through use of wood, feathers, and a wide range of pigments, including blood. Complex images, often made in sand, are a link with other worlds, creations of the mind. Songs, dances, elaborate rituals, all so much part of the Aborigines' cosmos, are yet all invisible in the archeological record. Even when material evidence of such rituals does remain—consider the prehistoric images that were carved or painted on rock shelters toward the end of the last Ice Age—they are stripped of much of their meaning in the absence of the context in which they were created.

A Pattern of Chance?

The shape of the human phylogenetic tree is not common in the history of life. Any group, or clade, that slowly slips into extinction must inevitably reach a stage with only one existing species, of course: a statistical requirement. But the persistence of a single successful lineage is rare, a distinction *Homo sapiens* shares with the aardvark. Although it is quite possible to conceive of two or more coexisting aardvark species, the coexistence of more than one species like us is beyond imagination. The trappings of culture and the world humans create through language, consciousness, and the force of mythology seem so powerful and all-enveloping as to exclude the possibility of sharing such an adaptive niche with another like species. Our unusual phylogenetic tree therefore may be a product of the type of animal we are. On the other hand, it may merely be a chance product of history that we erroneously imbue with significance: modern humans *are* the sole representatives of the hominid lineage, and as a result its members lack the imagination to think of themselves otherwise. This issue is an inescapable part of any contemplation of the origin of modern humans, whether dealt with implicitly or explicitly.

A Shift in Emphasis

In the early days of the twentieth century, anthropologists concerned themselves principally with the relatively recent events of human prehistory, which included the fate of the Neanderthals and the evolution of modern humans. The reason was simple: before 1925 no fossils of early human ancestors were known. Even when such fossils were discovered—by Raymond Dart in South Africa—a further two decades passed before the anthropological establishment would accept them as genuine members of the human

Raymond Dart, shown here in 1978 with the Taung skull, lower jaw, and natural brain cast—the remnants of an ape-like human ancestor that lived 1 to 2 million years ago—and the February 1925 issue of Nature, *which carried the announcement of the discovery.*

Mary Leakey established African archeology as a systematic study through her decades of careful work at Olduvai Gorge, Tanzania, beginning in the 1930s.

Louis Leakey's passion for the search for human origins in Africa paid off when his wife, Mary, found the "Zinjanthropus" cranium at Olduvai Gorge in 1959; many more fossil discoveries quickly followed.

family. There is no question that anthropologists were interested in finding the earliest representatives of the human family; but they were not prepared for what such creatures might look like (very much like apes) nor for where they would be found (Africa).

Beginning in the 1950s, a shift of attention occurred, both in interest and geography. The names of Raymond Dart and Robert Broom, and then of Louis and Mary Leakey, leapt to prominence as these legends of anthropology unearthed apelike fossils that

nevertheless clearly were early members of the human lineage, known formally as the family Hominidae, or hominids. South Africa, then East Africa, became the anthropological Mecca for headline-grabbing discoveries. The search for "the earliest" was on: the earliest stone tools, the earliest bipedal ape, the earliest member of the genus *Homo*.

The names of South African caves of Sterkfontein and Swartkrans became famous, as did Olduvai Gorge in Tanzania. Later, Richard Leakey, son of

Louis and Mary, became involved in the quest, with important discoveries east of Lake Turkana in northern Kenya. Even more spectacular were the fossils unearthed in the Hadar region of Ethiopia, from a joint American–French expedition run by Donald Johanson and Maurice Taieb. The famous Lucy skeleton came from there, as did the First Family, a collection of more than 200 fossil fragments representing at least 15 individuals from one location, all among the oldest hominids known. Most evocative of all was the trail of footprints, 3.7 million years old, discovered by Mary Leakey and her colleagues at the site of Laetoli, 20 miles southwest of Olduvai Gorge.

The 1960s and 1970s were the decades of "the earliest" in anthropology, and a very productive time it was. No doubt the picture developed from this work is still incomplete: more details of the pattern remain to be discovered, more branches to be affixed to the tree, more insight into the process and biology of this aspect of hominid history. Nevertheless, the two decades of anthropological fixation on East Africa and on the early events in human evolutionary history were scientifically justified and professionally satisfying to the protagonists.

Recently, however, anthropological attention has turned once again to the origin of modern humans and the fate of the Neanderthals. In each of the past several years at least half a dozen scientific conferences have been held on the subject, a sign of scholarly interest that goes beyond a simple decision to look again at an old problem. Fervor is hardly too strong a word to describe the renewed attention, and to judge from the pitch of debate, feelings run as high as they have ever done—a sure sign that something of special interest is going on.

The relative lack of spectacular fossil discoveries in recent years from the early part of the human evolutionary record unquestionably offered an opportunity for a new topic to take center stage in the debate over human prehistory. But the origin of modern humans did not assume this position simply by default: it was

thrust there by two new lines of evidence, one traditional, the other less so. The first concerned newly established dates for certain key fossils in the Levant, the Near East. The less traditional evidence was genetic and involved surveys of certain genes in modern populations as a way of reconstructing their past histories. In both cases, the results appeared to point to a radical interpretation of the origin of modern humans, one that turned against prevailing notions. These two lines of evidence were not isolated, however; other developments had been occurring as well, and these prepared the intellectual ground and contributed to the force of the new arguments. But the fossil redating and genetic evidence caught the imagination, particularly of the popular press.

One newspaper greeted the redating announcement with this proclamation: "Evolutionary theories were turned back to front this week when scientists claimed modern humans existed before Neanderthal Stone Age cavemen." A little exaggerated, perhaps, but even scholarly reaction was not entirely unrestrained. One anthropologist, commenting in the Brit-

This trail of footprints, discovered at Laetoli, Tanzania in 1974, were made by hominids about 3.75 million years ago. The prints reveal that hominids were habitual striding bipeds.

ish journal *Nature*, said that the new dates "[stand] the conventional evolutionary sequence on its head."

The genetic results found their way onto the cover of *Newsweek*, with the headline: "The search for Adam and Eve." The name Eve became intimately associated with certain genetic evidence, because by one reading of the data modern humans can trace their origins back to a single female who lived in Africa some 150,000 years ago. Perhaps it was inevitable that, because of the biblical allusion, there would

The Mitochondrial Eve hypothesis reached public notoriety in early 1987 with this cover of Newsweek.

be confusion over the proper interpretation of this data in both popular and scientific circles. Nevertheless, the force of the new genetic evidence—and its correct interpretation—has been powerful enough to demand serious attention.

Frustratingly Sparse Records

As we explore the currently developing notions of the origin of modern humans, many themes will emerge beyond the dominant one of pattern, the shape of the phylogenetic tree. These include the evolution of language, of consciousness, of subsistence and technology—the elements of behavior and mind that make us human. We will begin, however, with the prelude to *Homo sapiens*, the history of our ancestors as they first entered the novel niche of the bipedal ape and then exploited the subsequent adaptive radiation, leading eventually, but not inevitably, to us.

As the story of the search for modern human origins unfolds in these pages it will become evident that much remains to be learned and that considerable differences of opinion exist over available evidence. Differences of interpretation are perhaps inescapable over the elusive evidence of the emergence of language, consciousness, and ritual. But unresolved problems are not confined to those aspects of human evolution that make little tangible impression on the prehistoric record. Significant conflicts of interpretation exist over fossils and stone tools, as well. There are two principal reasons for this.

First, because human fossils often achieve public notoriety—sometimes under names such as Peking Man, Java Man, and so on—an impression builds in the public mind of an extensive fossil record. In fact, the human fossil record is sparse, even in the later period that is the focus of this book. The fossils represent a source of data of past populations. Faced with limited data, scientists are unable to reach unequivocal conclusions. Second, although fossils are one source of data on which anthropologists work, they do not represent unambiguous scientific facts as do physical measurements of temperature or weight or chemical composition. The shape of an anatomical feature may be established precisely, but its evolutionary relationship to similar features in other individuals must be inferred. Although anthropologists strive for objective methods by which to reach such inferences, it is not always possible, and subjective differences of opinion emerge.

Similar problems apply to the archeological record. Although the European record is relatively rich, elsewhere in the Old World it is poor. And because the form of artifacts is influenced by many factors—including the nature of the raw material, specific functional requirements, and local styles—determining relationships between different assemblages may be problematic. As challenging is the attempt to link changes in intellectual capacity with changes in artifact assemblages. The limited availability of material for study combines with the absence of objective criteria and leads to differences of professional opinion.

The story of the search as related here, therefore, is as much about how anthropologists attempt to know the past as it is about what they do learn. The urge to uncover events in human prehistory is strong in professionals and amateurs alike, as is the frustration at unanswered questions. There is no mistaking the intensity and excitement of the quest, as science year by year moves ever closer to its goal.

1

PRELUDE TO *HOMO SAPIENS*

No one is more strongly convinced than I am of the vastness
of the gulf between . . . man and the brutes . . . for, he alone
possesses the marvelous endowment of intelligible and rational
speech [and] . . . stands raised upon it as on a mountain top,
far above the level of his humble fellows, and transfigured
from his grosser nature by reflecting, here and there, a ray
from the infinite source of truth.

— *Thomas Henry Huxley, 1863*

Huxley, Charles Darwin's friend and champion, was correct to point
to the special nature of *Homo sapiens*; but it was not always so. In
human prehistory, "the gulf between . . . man and the brutes" did
not exist. It is a product of the evolutionary process—of the accre-
tion over time of powerful adaptations by a bipedal ape. Nor was
the emergence of this gulf by any means inevitable. Evolution is a
process of the moment, responsive to prevailing circumstances—
specifically those of climate and environment. If conditions at any
particular point in the past had been different, subsequent evolutionary history would have been different, too.

Skull of Homo habilis *(KNMER 1470, from Koobi Fora,
Kenya) and bones of the right foot of* Homo habilis
*(OH8) from Olduvai Gorge, Tanzania. (Both specimens
are shown as casts.)*

There is every reason to believe, for example, that the fragmentation of rain forest in East Africa, already under way 10 million years ago, was instrumental in the initial evolution of the bipedal ape group called the hominids. Earlier, dense tropical forest had formed a belt across Africa, providing an extensive habitat suitable for apes. This belt of vegetation was interrupted by events deep beneath the continental crust, an aspect of plate tectonics. Following a line that runs from Turkey through Israel and the Red Sea, snaking down through Ethiopia, Kenya, and Tanzania and finally into Mozambique, two tectonic plates are slowly moving apart. The geological consequences have been spectacular. First, under pressure from rising magma, two huge blisters, known as the Kenyan Dome and the Ethiopian Dome, were raised in the continental rock, each more than 6000 feet high. Second, just as the Red Sea is widening by about 1 millimeter per year as the plates move apart, the continental rock in the line just mentioned was put under great strain and the plates beneath separated and eventually collapsed (faulted), producing the Great Rift Valley. Third, in addition to a complex pattern of geological faulting, volcanic activity broke out all along the valley, which soon became scattered with alkaline lakes.

As a result of all this tectonically driven activity, the land to the east of the Rift Valley was thrown into rain shadow, and the range of different environmental conditions from the arid desert floor to the cool highlands generated a tremendous mosaic of habitats where previously there had been one, that of dense tropical forest. These new habitats represented evolutionary opportunity. Almost certainly the new, more open habitats played a key role in the initial evolution of hominids from a species of forest ape. Had these plate-tectonic events not occurred, or occurred differently, bipedal apes might never have evolved. Similarly, some scholars believe that global cooling about 2.6 million years ago prompted further alteration of habitats, causing another burst of evolutionary change. South African paleontologist Elisabeth Vrba, now at Yale University, has shown that many new antelope species arose at this time. New hominid species evolved too, including the first of the large-brained bipedal apes, the genus *Homo*. Without that environmental event, today's bipedal apes might never have achieved brains the size of *Homo sapiens*'s. Huxley's gulf would not have come into being.

Like all species, human beings are the product of historical contingency. When we examine, in the following pages, the prelude to *Homo sapiens*, we are looking at the history of a species: the unfolding phylogenetic pattern of its ancestors. Although reconstruction may tempt us to view each step as preparation for the next, the pattern observed is only one of many that were possible. Retrospect lends it the appearance of inevitability, but the final pattern that we will attempt to trace is the result of environmental circumstances as they happened to have occurred.

What kind of animal was the immediate forebear of modern humans? To answer this question, we must consider the overall pattern of major adaptations—anatomical and cognitive—that effectively transformed an ape into a human. We will also take into account the origin and development of technology, changes in social structure, and the evolution of novel modes of subsistence, such as the hunting and gathering lifeway. Our exploration will begin with a comparison of ape and human anatomy, benchmarks for comparing other creatures in our lineage.

Apes and Humans Compared

Perhaps the most obvious physical difference between humans and the African apes is in posture and locomotion. Humans stand and walk on two feet, whereas the apes are quadrupeds, employing the so-called knuckle-walking mode of locomotion. Chimps and gorillas are able to stand and walk on two legs,

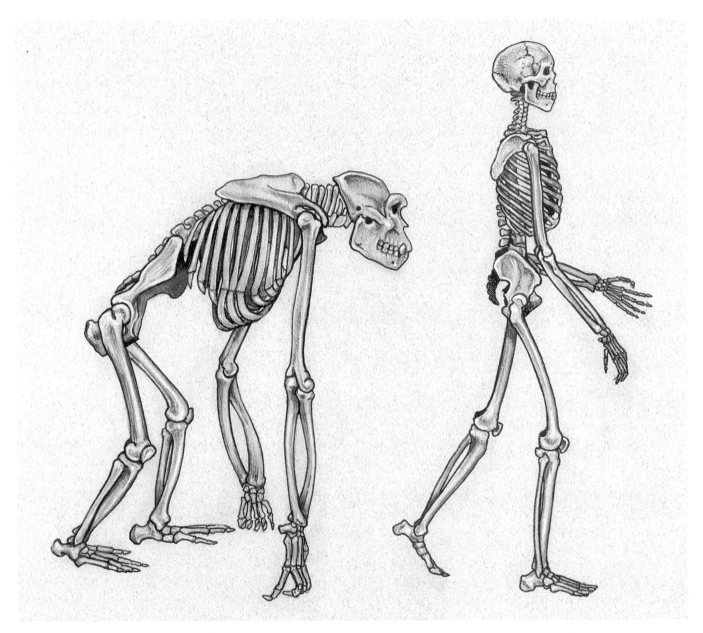

A comparison of posture and skeletal proportions in apes and humans reveals different adaptations. Apes move by a form of locomotion known as knuckle-walking, whereas humans are habitual bipeds. In humans the legs are long relative to the arms, while the reverse is true in apes. The ape pelvis is long and narrow, compared to a short, wide pelvis in humans. The lumbar region is longer in humans than in apes.

but this is an awkward, transitory posture for them. Much of the anatomical distinction between humans and apes relates to this difference in locomotion, although, as we shall see, modern humans are specialist bipeds compared with earlier hominids.

In humans the arms are short, the legs long, and the trunk relatively slender compared with apes. Overall, humans are slightly built for their height, whereas apes may be described as bulky or stocky. In both apes and humans the rib cage is broad from side to side and relatively flat from front to back, a configuration that contrasts with the deep rib cage of monkeys. Nevertheless, the human rib cage is somewhat barrel-shaped compared with the conical or funnel-shape in chimpanzees and gorillas. Humans and apes share similar pectoral girdle (shoulder) anatomy, affording considerable movement of the arms: an adaptation to climbing in the apes and at least partly a mark of ape heritage in humans. Nevertheless, the pectoral girdle is situated higher in apes than in humans, a position that restricts the kind of arm-swinging characteristic of striding humans. The high shoulders of apes also prevent the very deep breathing of which humans are capable: an adaptation important during sustained running.

The lumbar region in humans is longer than in apes, while the pelvis is shorter. One consequence is that apes have no waist, which restricts their flexibility and the kind of lateral movements that humans employ while running. The general abdominal structure in apes makes them rather pot-bellied in contrast with the build of humans. The combination of anatomical features throughout the trunk implies an efficient adaptation to running in humans versus climbing in apes. In becoming specialized bipeds, however, humans have retained the flexibility of other modes of locomotion. As the British biologist J. B. S. Haldane once noted, only humans can swim a mile, run 20 miles, then climb a tree.

Human legs are not only relatively longer than those of apes, but their configuration is different.

Most important, the thigh bone angles in toward the knee, placing the feet close together directly under the body's center of gravity. In apes there is very little angle to the knee, planting the feet far apart: a position that contributes to the awkward waddle when apes walk bipedally. Human feet are flat platforms, springboards for the push-off during bipedal walking and running. In apes, by contrast, the foot bones are somewhat curved and the great toe is splayed outwards—all adaptations to climbing. Similarly, ape hands are much more curved than human hands; they also lack the opposable thumb characteristic of the precision grip in humans. Human fingertips are flatter and broader than apes', a consequence of the evolution of finger pads packed with sensitive nerve endings.

Now to the head. The most important difference is the size of the brain, some 400 ccs in gorillas and chimpanzees compared with 1350 ccs in humans. If humans were apes of the same body size, our brain would be one-third of its present volume. In some apes, particularly gorillas, a bony crest runs like a keel from the front to the back of the cranium, an anchor for the large muscles that power the lower jaw. The ape's face juts out considerably, a configuration known as prognathic, as compared with the flat face of humans. The human face shape is the result of the tooth rows having been "tucked under" the cranium, a position that increases grinding efficiency. Tooth rows in apes are somewhat U-shaped; in humans, they form a parabolic arch. One dramatic difference here is the size of the canine teeth, which can be enormous in apes, projecting well beyond the level of the tooth row; human canines barely exceed the height of adjacent teeth. In apes a gap (diastema) is present between these large canines and the adjacent incisors, into which the canine from the opposing jaw may slip when the tooth rows are brought together. No diastema appears in humans.

Ape dentition is designed for processing fruits. The incisors are therefore large and somewhat forward-projecting, whereas the cheek teeth (premolars and molars) are relatively small, with high cusps capable of slicing fruit. Ape teeth are coated with a thin layer of enamel. In humans the incisors and canines, all small, form a single slicing row, whereas the relatively large cheek teeth have low cusps that wear flat to become an efficient grinding surface. Human teeth possess a thick layer of enamel. The human dentition is a machine with many functions, capable of processing both meat and plant foods.

A Sparse Early Record

African apes and humans may be thought of as the starting- and end points of the hominid group, with two caveats. First, present-day African apes are not our ancestors, any more than humans are their ancestors: rather, humans and African apes share a common ancestor, from which each lineage must have varied through evolution. Nevertheless, from what we can discern from a very incomplete fossil record, the African apes can be used as a model of apedom, even though their knuckle-walking may possibly be a specialized adaptation. Second, the current representative of the hominid group—*Homo sapiens*—should not be thought of as the final product toward which earlier evolutionary changes were moving. It is by no means the case that all hominid morphological characteristics changed consistently in the direction seen in modern humans. Bearing these two warnings in mind, it is now possible to look at the adaptive radiation of hominids, using "apeness" and "humanness" as extremely rough measures of certain changes, providing us with key areas of comparison: posture, overall body proportions, dentition, and brain size.

If hominids followed the typical pattern of mammalian evolution, an adaptive radiation from the founding species of the group would be expected to develop over time. This, as far as we can tell, is what

occurred; but the picture is very incomplete, particularly at the earliest stages. Although molecular evidence—which we will examine next—indicates that the hominid group was established about 7.5 million years ago, the earliest putative hominid fossil (a small fragment of cranium) is just 5 million years old. Indeed, the evidence is extremely fragmentary until we reach a little in excess of 3.5 million years old: the jaw fragments and footprint trail at Laetoli in Tanzania. The record begins to improve substantially toward the present, with material from the Hadar region of Ethiopia, the Lake Turkana sites in northern Kenya, Olduvai Gorge in Tanzania, and various South African cave sites. Virtually all hominid fossil remains earlier than about 1 million years have been found in Africa.

One reason why the earliest part of the hominid fossil record is so sparse is the simple, though frustrating, limitation of geological exposures of the appropriate age. The virtual absence of hominid fossils older than about 1 million years outside the African continent, however, is considered a genuine reflection of history. Hominids did not, apparently, move beyond their tropical range of origin until that time.

Molecular Anthropology Makes a Big Impact

One of the most decisive developments in twentieth-century anthropology occurred in the early 1960s. It derived not from the fossilized bones of long-extinct species but from the blood of living animals. In anthropology texts of the time, the African great apes (chimpanzee and gorilla) and the Asian great ape (the orangutan) were grouped together, implying a close evolutionary relationship, while humans were placed evolutionarily distant. Biologically speaking, chimpanzees are known by their genus and species name *Pan troglodytes*, gorillas by *Gorilla gorilla*, and the orang-

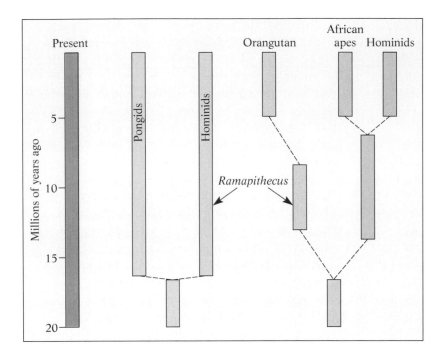

Views of ape/human relations changed dramatically in the late 1970s. In the earlier hypothesis (left) humans (hominids) and pongids (apes) were thought to have diverged by 15 million years ago, making Ramapithecus *a putative human ancestor. By 1980 hominid origins were seen as being closer to 5 million years ago, and* Ramapithecus *was judged to be part of a group ancestral to orangutans.*

utan by *Pongo pygmaeus*. Traditionally, the three apes were placed in the biological family Pongidae, informally known as the pongids. Humans, *Homo sapiens*, according to the traditional view, are the sole occupants of the biological family Hominidae, or hominids. Morris Goodman, a biochemist at Wayne State University, demonstrated that this grouping is probably incorrect: the African great apes and humans are each others' closest relatives, while the orangutan is the genetic outsider. This genetic relationship, Goodman suggested, should be reflected in the formal classification too: The African apes and humans should share the family Hominidae, while the orangutan should be the only member of the Pongidae.

In his experiments Goodman used a simple but effective test as a measure of genetic relatedness, based on the strength of an antibody response to the protein albumin from the blood of apes and humans. If the structure of albumin from two species was

identical, the strength of the response should also be identical; the more different the structure of one protein from another, the greater the difference in the strength of the response. The assumption underlying this and all such tests is that similarity of protein structure indicates close genetic relationship, while differences in structure indicate genetic distance. The results Goodman obtained with this method challenged the prevailing anthropological hypothesis that the three great apes were evolutionarily more closely related to each other than either was to humans.

Darwin, Huxley, and the German biologist Ernst Haeckel had come to the same conclusion as Goodman a century earlier. They based their conclusions on comparative anatomy: chimpanzees, gorillas, and humans share certain important anatomical features, while orangutans uniquely possess certain other features. But these conclusions become superseded as the decades passed, primarily because of an emphasis later researchers placed on shared adaptations: the way of life of the three great apes seemed to be very similar, while the human adaptation was clearly different. Shared adaptation among the apes was taken as an indication of shared ancestry—incorrectly, as it turned out. With Goodman's results on the comparison of serum proteins, which gets much closer to the fundamental material of inheritance, the science of molecular anthropology was effectively born—and the Darwinian view reinstated. Anthropologists had long argued about the identity of the immediate ancestor of humans, and from the genetic evidence it had now become clear. It was some kind of ape—not a chimpanzee, but something from which the chimp and gorilla also descended.

In addition to providing information about genetic relatedness, molecular anthropology has also been used to set a time frame for human evolutionary history. In the late 1960s, when most anthropologists were arguing that the first hominid evolved some 15 million—perhaps even 30 million—years ago, University of California, Berkeley, biochemists Allan Wilson and Vincent Sarich produced a very different answer. Using a technique like Goodman's, but applying the notion of a molecular clock, they announced that human evolution was a mere 5 million years old. Wilson and Sarich arrived at this figure by first calibrating the molecular clock: the rate of accumulation of genetic change through time. They measured the genetic distance between humans and Old World monkeys, from which, according to fossil evidence, we separated about 30 million years ago. They then noted that the genetic distance between humans and apes was one sixth that between humans and Old World monkeys, and therefore concluded that humans and apes separated from each other 5 million years ago (one sixth of 30 million years).

When Wilson and Sarich published this conclusion in 1967 in *Science*, they received sharp criticism from molecular biologists and anthropologists, particularly from the latter group. Many argued that there was no theoretical reason why the accumulation of genetic change should be steady through time—that

the molecular clock should, in effect, tick regularly. The same point continues to be argued today with respect to molecular evolution in general: the subject is both complicated and controversial. However, it is possible to determine, as Wilson and Sarich did more than 20 years ago, when the clock ticks regularly and when it does not. In the case of genetic change of albumin in the great apes and humans, the ticking had been regular, they concluded.

For a decade and a half, Wilson and Sarich's suggested date for the origin of hominids continued to be rejected by most anthropologists. As time passed, new genetic techniques became available with which to test the conclusion. The original work, which depended on antibody reaction to differences in protein structure, was effectively several steps removed from the ultimate source of genetic information, DNA itself. New techniques approached ever closer to that source: first with the ability to read the sequence of amino acids that formed the primary structure of proteins; then, at the level of DNA by first mapping certain markers and then reading the sequence of nucleotide bases that form its fundamental structure. Another technique, known as DNA hybridization, effectively matches up the entire complement of DNA from one species against that of another, rather than examining single genes or sections of genes as the other techniques do. As each development came along, the answer was the same: Wilson and Sarich had been correct in saying that the origin of hominids was a recent evolutionary event. The date for that event was perhaps a little earlier than they had originally proposed, around 7.5 million years ago. However, most anthropologists now accept this date as being consistent with the fossil record.

When the genetic intimacy between humans and the African apes had been assimilated during the 1970s and early 1980s, most scholars assumed that gorillas and chimpanzees were most closely related to each other, with humans separate; this seemed as obvious to the untrained eye making observations at the zoo as it did to experts doing comparative

*M*olecular *Anthropology*

The term molecular anthropology was coined in 1962 by Emile Zuckerkandl of the Linus Pauling Institute, California, at a meeting of the Wenner-Gren Foundation for Anthropological Research in Burg Wartenstein, Austria. Zuckerkandl, in collaboration with Pauling, had earlier developed the notion of reconstructing phylogenetic relationships among species based on comparison of genetic similarities and differences. The 1962 meeting was titled "Classification and Human Evolution," and Zuckerkandl was pushing hard his concept that molecules can be as revealing about evolutionary relationships as the more traditional anatomical evidence—in some cases perhaps even more so. All the big names in evolutionary biology and physical anthropology were there: George Gaylord Simpson, Ernst Mayr, Theodosius Dobzhansky, Louis Leakey, Sherwood Washburn—an impressive list. The giants of the traditional approach were interested, but skeptical. At the end of it all, Dobzhansky said to Zuckerkandl:

"Maybe in twenty years time you will be able to say, 'I was right.'"

Dobzhansky's prediction was correct. Phylogentic reconstruction based on genetic evidence has become an important tool in biology during the past decade—not least in anthropology, where it has been used to address three major questions: the origin of the hominid family; the origin of modern humans; and the timing and pattern of human occupation of the Americas. In theory, the technique should be able to produce unequivocal answers to these questions. The molecular context of mutational change is more complicated than was once imagined, however, thus making the interpretation of genetic information less straightforward than had been hoped. Evidence from molecular anthropology may be taken as complementary to more conventional data from paleontology and archeology: ultimately, the three lines of evidence should tell the same story.

The technique of molecular anthropology has its origins much earlier. In the early 1900s George Henry Falkner Nuttall, a professor of chemistry at Cambridge University who had studied with the great German biologist Paul Ehrlich, developed an immunological method for investigating genetic relationships among species. He compared the reaction of blood from species in question to a range of antisera: closely related species react similarly to the same antiserum. In a large series of experiments he investigated certain primate species, including humans. He had this to say in a lecture at the London School of Tropical Medicine, 28 November 1901: "If we accept the degree of blood reaction as an index of blood relationship within the Anthropoidea, then we find that the Old World apes are more closely allied to man than are the New World apes, and this is exactly in accordance with the opinion expressed by Darwin."

Nuttall had carried out his pioneering work just three decades after Darwin published his *Descent of Man*, in which he speculated on the implications of the close anatomical relationship between humans and African apes. But six decades passed before molecular anthropology was taken further, by Morris Goodman of Wayne State University. Although Goodman had inferred the shape of the phylogenetic tree from his own immunological data—in this case a three-pronged fork—he put no scale

Allan Wilson did much to further the development of molecular anthropology.

on the timing of evolutionary events. The idea of using the extent of mutation as an indication of the passage of time—a molecular clock—was in the air at the time, though not well formulated. Emanuel Margoliash made the concept explicit in 1963:

"If elapsed time is the main variable determining the number of accumulated substitutions, it should be possible to estimate roughly the period at which two lines of evolution leading to any two species had diverged." In other words, a molecular clock exists if the accumulation of mutations occurs on average at a steady rate. It was Vincent Sarich and Allan Wilson, not Goodman, who applied the molecular clock to the human–ape relationship.

Before applying the notion of a clock, Sarich and Wilson had to demonstrate that it worked. This they did with the rate test, which they devised. If two species (A and B) are recently diverged from each other but share a common ancestor with a third species (C), and if the average rate of mutation in lines leading to each of the species is the same, then the total amount of genetic difference between species A and species C will be the same as between species B and species C. Deviations from the average rate in any of these lines will lead to different results for the A-to-C and B-to-C comparisons.

Sarich and Wilson attempted to publish a paper on the rate test in *Science* before they announced the molecular clock results, but the manuscript was turned down as demonstrating nothing new or interesting. It was published in the fall of 1967 in the *Proceedings of the National Academy of Science*, a journal not frequently read by anthropologists. When later in the same year the two biochemists published "Immunological Time Scale for Hominoid Evolution" in *Science*, in which they showed that humans and apes had diverged 5 million years ago—not the 15 million or more years then held by anthropologists—one criticism they faced was that they had *assumed* a constant rate for the accumulation of mutations. Sarich recently mused: "One wonders if the continual accusation that we somehow 'assumed' the clock and imposed it on the data stems in large part from this original misperception on the part of the reviewers *Science* used, as the *Proceedings* article has virtually never been referred to or critiqued." The often acerbic debate over the validity of Sarich and Wilson's genetic data and their interpretation marked the first major controversy in molecular anthropology, one that surely influences the tenor of similar debates today.

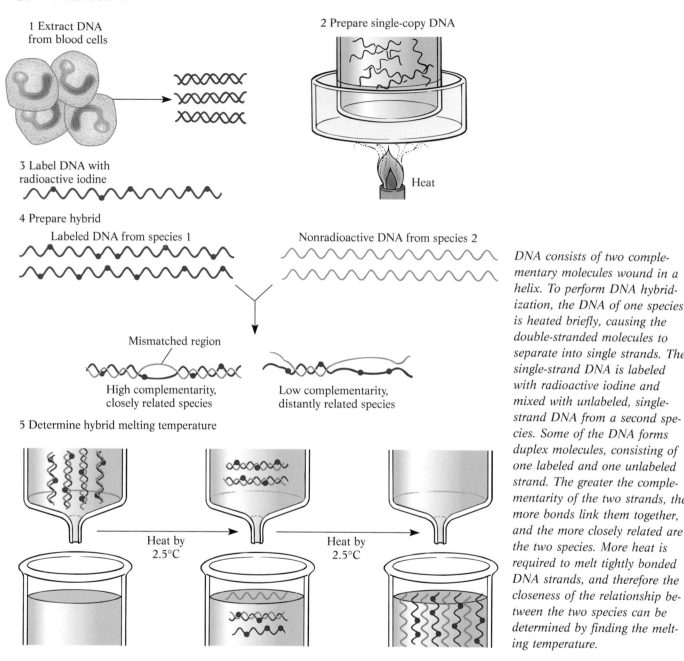

1 Extract DNA from blood cells

2 Prepare single-copy DNA

Heat

3 Label DNA with radioactive iodine

4 Prepare hybrid

Labeled DNA from species 1

Nonradioactive DNA from species 2

Mismatched region

High complementarity, closely related species

Low complementarity, distantly related species

5 Determine hybrid melting temperature

Heat by 2.5°C

Heat by 2.5°C

DNA consists of two complementary molecules wound in a helix. To perform DNA hybridization, the DNA of one species is heated briefly, causing the double-stranded molecules to separate into single strands. The single-strand DNA is labeled with radioactive iodine and mixed with unlabeled, single-strand DNA from a second species. Some of the DNA forms duplex molecules, consisting of one labeled and one unlabeled strand. The greater the complementarity of the two strands, the more bonds link them together, and the more closely related are the two species. More heat is required to melt tightly bonded DNA strands, and therefore the closeness of the relationship between the two species can be determined by finding the melting temperature.

anatomy in the laboratory. In 1984, however, Charles Sibley and Jon Alquist, then at Yale, published DNA hybridization evidence supporting the counterintuitive grouping: that humans and chimpanzees were each other's closest relatives, with gorillas separate. Since then the same conclusion has been inferred from

DNA mapping and sequencing evidence. If true—and the notion is not yet universally accepted—there are interesting implications for the mode of locomotion of the common ancestor of the African apes and humans: more likely than not, it was a knuckle-walker.

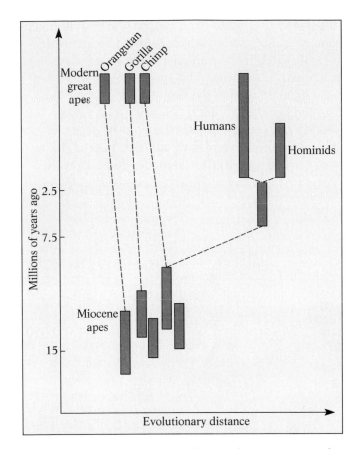

An assessment of evolutionary distance between apes and humans may be based on adaptation. Although humans are genetically close to the African great apes, in terms of adaptation humans may be considered distant from the apes.

The argument is in a sense statistical. If the common ancestor was not a knuckle-walker, and if chimpanzees and gorillas arose separately (as the new genetic evidence suggests), then knuckle-walking evolved independently in the two lineages. This is much less likely than the assumption that the common ancestor was a knuckle-walker and that humans adopted a different mode of locomotion. Since no strong evidence, however, in hominid anatomy indicates a knuckle-walking ancestry, the argument remains unsettled.

Here, then, we have the framework for the prelude to *Homo sapiens*, anatomically and temporally. Under the selection pressure of changing environmental circumstances—the fragmentation of forest cover some 7.5 million years ago—an apelike creature evolved a novel adaptation, namely, bipedal locomotion. We will trace the adaptive radiation of bipedal apes as it occurred over the next 7 million years, bringing us to the threshold of *Homo sapiens*, 500 thousand years ago.

Energetics of Bipedalism

The primary hominid adaptation—the evolutionary innovation that established the group—was bipedalism. Many hypotheses have been advanced over the decades as explanations of this event, some involving specifically human elements such as culture, hunting, and stone-tool making. However, as the earliest evidence of stone-tool making occurs only 2.5 million years ago, long after hominids became established, other explanations must be sought. The most promising line of argument focuses on foraging. The thinning out of forests that was under way in East Africa by 10 million years ago would have gradually disrupted traditional ape habitats. With food patches more widely dispersed, selection pressure would favor more efficient locomotion over the terrain.

True quadrupedality, as seen in dogs, horses, and other such animals, has long been known to be more energy efficient than human bipedality: we burn more calories than a dog in covering the same distance at speed, and our top speed is relatively slow. Many observers therefore assumed that the evolution of human bipedality could not have been for reasons of locomotor efficiency. A decade ago University of California, Davis, anthropologists Peter Rodman and Henry McHenry pointed out what should have been obvious: human locomotor efficiency should be compared with that of an ape, not a dog. Chimpanzees, in fact, are about 50 percent less energy efficient than humans when walking on the ground, whether they are bipedal or quadrupedal. Therefore, said Rodman and McHenry, there is no energy barrier in going from apelike quadrupedalism to human bipedalism; indeed, there is benefit to be gained for an animal that covers large distances on the ground. They suggested that hominids evolved bipedally as an adaptation to foraging for an apelike diet that had become distributed over more open countryside than is suited to apelike locomotion, which tends to be arboreal.

If bipedality was an adaptation for efficient locomotion, the first hominid species might well have been hominid only in the way it stood and moved. Traveling between broadly scattered food patches but seeking and eating the same items as its ancestors, this first hominid species would have been under no selection pressure to change its dentition. However, in the first known hominid species—*Australopithecus*

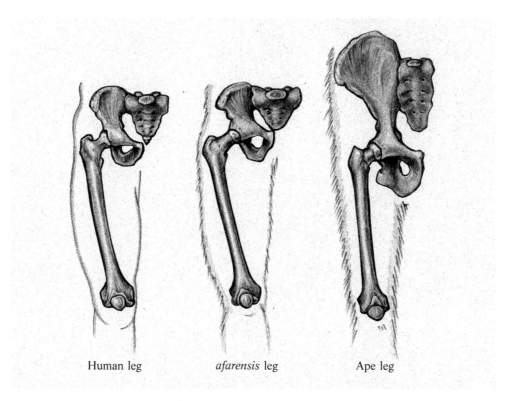

Human leg   *afarensis* leg   Ape leg

*Lower limb anatomy of modern humans, an early hominid (*Australopithecus afarensis*), and apes reflects styles of locomotion. The angle between the thigh bone (femur) and the knee, known as the valgus angle, is critical to efficient bipedalism. The significant valgus angle in humans and early hominids allows the feet to be placed under the midline of the body while striding. The absence of such an angle forces an ape to place its feet wide of the midline, thus to "waddle" while walking bipedally.*

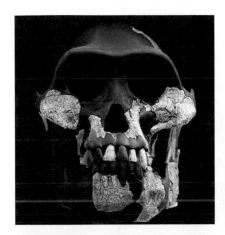

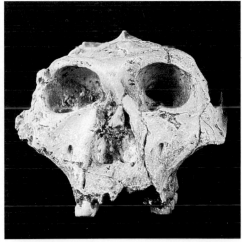

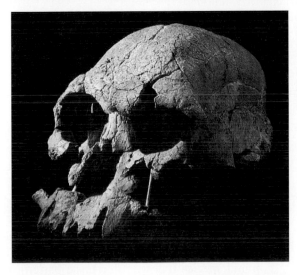

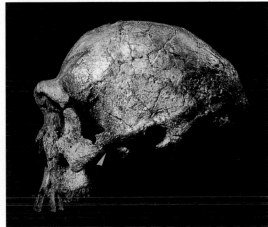

Upper left: Australopithecus afarensis; middle left: A. boisei, *seen in side view, specimen 406, from Koobi Fora, Kenya (age: 1.6 million years)*; lower left: Homo habilis, *specimen 1470, from Koobi Fora, Kenya (age: 1.9 million years)*; upper right: A. robustus; lower right: Homo erectus, *seen in side view, specimen 3733, from Koobi Fora, Kenya (age: 1.6 million years)*.

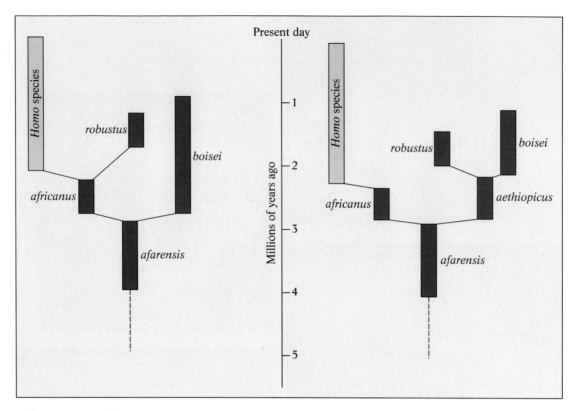

Different views of hominid phylogeny: The principal division based on adaptation is between large-brained, small cheek-toothed hominids (the Homo *species) and the small-brained, large cheek-toothed species of* Australopithecus *(all those species named here). Differences of opinion exist over the relationship of various of the more robust species, namely,* robustus, boisei, *and* aethiopicus.

afarensis, dated to 3.75 million years ago at most—the dentition differs from an ape's. It is humanlike to the extent of displaying relatively small canines, a thick enamel layer, and cheek teeth better designed for grinding tough material than for dealing with fleshy fruit. There are, on the other hand, many primitive—that is, apelike—features about it, including the occasional presence of a diastema, somewhat protruding incisors, and certain characters in the premolars. The tooth row shows neither the typical U-shape of apes nor the parabolic arch of modern humans: it is more V-shaped. The overall appearance of the head is apelike, with a small brain and prognathic jaw. With fossilized leg and pelvic bones indicating a strong adaptation to upright walking, this early hominid was very much a bipedal ape, though not completely so. The bones of the feet and hands were slightly curved, possibly indicating a partly arboreal habit. And though the hands were modern in general shape, the fingers had not yet developed broad pads.

Until a little more than 2 million years ago, all hominid species followed this basic pattern: they were small-brained, large cheek-toothed, bipedal apes, but the curvature of the hand and foot bones was not pronounced, as far as can be determined. By this time adaptive radiation had clearly moved the bipedal ape away from an apelike diet, exploiting a much wider range of plant foods. In some species the adaptation to processing tough plant foods was extreme, the molars becoming enormous and the anterior teeth tiny. *Australopithecus robustus* and *Australopithecus boisei* are two examples of this pattern, and both display the bony keel along the top of the head. A less extreme variant was *Australopithecus africanus.*

The *Homo* Adaptation

Then a second shift occurred. This novel adaptation can be described as a large-brained, small cheek-toothed, bipedal ape—the beginning of the genus *Homo.* Although of the basic hominid pattern, its molars and premolars were smaller and anterior teeth larger. The earliest known species to follow this adaptation was *Homo habilis,* the oldest fossils of which are close to 2 million years old. *Homo erectus* followed, beginning about 1.7 million years ago. The new anatomy probably involved a dietary shift that

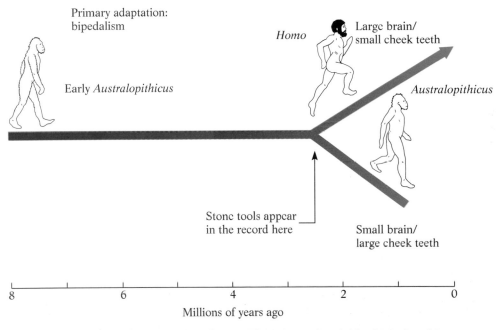

Divergent adaptations in hominid evolution: The primary hominid adaptation, bipedalism, arose somewhere between 8 and 5 million years ago. About 2.5 million years ago stone-tool making began to appear in the record. Shortly thereafter two forms of adaptation emerged: one was that of large-brained, small cheek-toothed species (Homo); *the second was a small-brained, large cheek-toothed adaptation* (Australopithecus, *only).*

Ramapithecus: The Ape That Was Not a Hominid

During the 1960s and 1970s the majority of anthropologists believed that the human family—the Hominidae—had arisen at least 15 million years ago. This conclusion was based on what was considered a primitive hominid, *Ramapithecus*, fragmentary fossil evidence of which had been found at sites in Europe, Asia, and Africa, the oldest of which was 15 million years and the youngest 8 million. In the early 1980s new fossil evidence from Turkey and Pakistan showed conclusively that *Ramapithecus* was not a hominid, merely a Miocene ape. This new evidence effectively removed the barrier between the established anthropological view and the biochemical evidence, which indicated that hominids first evolved a mere 5 million years ago.

Fossil fragments of *Ramapithecus* were first found in 1932 in northern India by G. Edward Lewis, a graduate student at Yale University. In a paper in 1934 he suggested that the fossils, two fragments of an upper jaw, were those of an early hominid. His conclusion was strongly criticized by Aleš Hrdlička of the Smithsonian Institution, a powerful figure in American anthropology, who argued that the animal was an ape. For almost three decades *Ramapithecus* was considered just one of many Miocene apes. In 1961 Elwyn Simons, then of Yale but now at Duke,

G. Edward Lewis (a self-portrait), in the Siwalik Hills, India, in 1932, the year he found the first specimen of Ramapithecus.

reexamined the fossils and declared that Lewis had been correct. Shortly thereafter Simons was joined at Yale by David Pilbeam, and the two men became the most prominent proponents of the *Ramapithecus*-as-hominid school.

Simons based his conclusion on the shape of the jaw and the configuration of the teeth. In modern apes, the tooth row forms a U-shape; in humans it is an arch. Although the two fossil fragments of upper jaw missed a central section and therefore did not join, Simons reconstructed the jaw and concluded that

it too was arched. Further, the overall shape of the jaw indicated that the animal's face had been short, as in humans, rather than projecting forward, as in apes. The most striking difference in the overall dentition between humans and apes is the canine: the canines in *Ramapithecus* are small, as in humans. The molar teeth are also humanlike, having low cusps adapted to grinding.

Additional humanlike characteristics were noted later, such as the thick enamel layer and apparent differential wear on the molars. Because human childhood is extended there is a considerable time gap between the eruption of the first, second, and third molars. As a result, the wear on the three teeth in a young adult is quite different, being most on the first molar and least on the third. Such differential wear is much less evident in apes. Although less marked than in humans, differential wear was apparent on the *Ramapithecus* molars.

Note that in these assessments of the status of *Ramapithecus*, the model for primitiveness was that of modern apes, specifically African apes. This was a common, and mistaken, assumption of the time. Fossil species are rarely identical to living species and may have unusual combinations of features.

In their descriptions of *Ramapithecus* Simons and Pilbeam suggested that it was probably bipedal, made tools, and hunted. These deductions were based on the Darwinian model of human origins, in which hominids were essentially cultural from the start. For instance, the reduction in the size of canines was explained as the result of tools and weapons taking over the role of the sharp teeth, just as Darwin had speculated in his *Descent of Man*.

In 1967 Vincent Sarich and Allan Wilson challenged anthropological orthodoxy: biochemical evidence indicated that the first hominid evolved only 5 million years ago, they said, and therefore *Ramapithecus* cannot be a hominid because it is too old. Most anthropologists felt that if *Ramapithecus* was to be ousted it should be as a result of conventional evidence, and that is what happened.

The first evidence concerned the shape of the jaw. In 1973 Peter Andrews of the Natural History Museum, London, and Alan Walker of the Johns Hopkins University published a reconstruction of an upper jaw allied to *Ramapithecus* (known by some at the time as *Kenya pithecus*). Contrary to previous interpretations, the jaw did not match the humanlike arch but was much more apelike. Three years later, in 1976, a member of Pilbeam's expedition in Pakistan discovered a *Ramapithecus* lower jaw, shaped like a truncated **V**. One pillar of the *Ramapithecus*-as-hominid hypothesis therefore crumbled. The original reconstruction of the 1932 jaw had clearly been in error.

Also in 1976 the neat dichotomy between thin and thick enamel began to break down. Several Miocene apes were discovered to have thick enamel, as does the orangutan. *Ramapithecus*'s thick enamel is therefore no indication of an advanced state or of hominid status. The use of African apes as a model of what was primitive was clearly mistaken. The most crucial fossil evidence, however, concerned a different animal, *Sivapithecus*, important examples of which were announced in 1980 and 1982, the first from Turkey, the second from Pakistan. *Ramapithecus* and *Sivapithecus* were considered closely related.

The conceptual context into which the new fossil evidence fit was important. Anthropologists had by this time accepted the notion that humans and African apes were closely related, while the Asian great ape (the orangutan) was more distant. Because the new specimens of *Sivapithecus* represented virtually complete skulls, more anatomical detail was available than previously. From this detail, particularly the shape of the eye orbits, midfacial region, and palate, it was clear that *Sivapithecus* was closely allied to the orangutan, perhaps part of a group ancestral to the living species. If *Sivapithecus* were closely related to the orangutan, then *Ramapithecus* must be, too (being itself closely related to *Sivapithecus*). As a relative of the orangutan *Ramapithecus* was automatically disqualified from human ancestry, since humans are related to African, not Asian, apes.

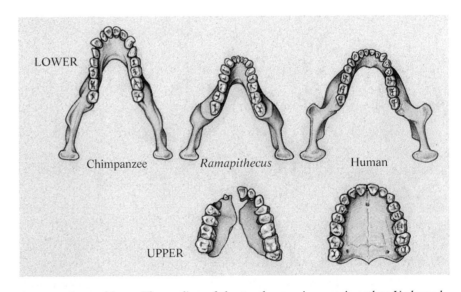

A comparison of jaws: The outline of the tooth rows in apes is rather U-shaped, while in humans it is parabolic. Reconstruction of an incomplete upper jaw of Ramapithecus in the 1960s led researchers to believe it followed the human form, and was therefore an early hominid. Later discoveries revealed that the jaw shape of Ramapithecus was distinct from that of apes and humans and belonged to a species of Miocene ape.

included meat. It is tempting to link the beginnings of stone-tool making to this adaptation—a link that would gain support should fossils of a *Homo* species one day be found as early as 2.5 million years, which is when stone tools first appear in the archeological record. With the emergence of *Homo*, the hominid group was at its most densely branched (accepting our ignorance of events earlier than about 3.75 million years ago). At least three species coexisted—maybe twice that number.

The adaptive shift that brought with it the large-brained, small cheek-toothed, bipedal ape was accompanied by other significant modifications in body shape and size. From the limited amount of skeletal evidence available so far, there is no question that by 2 million years ago all hominids were fully capable bipeds: the overall shape of the pelvis, the angle of the thigh bone, and the platform structure of the feet attest to that. Independent studies by Leslie Aiello of University College, London, and Peter Schmid of the Anthropological Institute, Zurich, have recently shown, however, that while the australopithecines were stockily built, earliest *Homo* was a much more slender creature. The size and shape of bones and the prominence of muscle attachments give clues to the build of these extinct creatures. Australopithecines, it appears, were built more like apes than like humans.

For comparison, a 6-foot-tall chimpanzee would weigh twice as much as a human of that height, and so would a 6-foot-tall australopithecine. A 6-foot-tall early *Homo*, by contrast, would be very humanlike in build, as can be seen in one of the most spectacular discoveries of the century, a 1.6-million-year-old *Homo erectus* youth from west of Lake Turkana. Although powerfully muscled, the so-called Turkana boy was tall and slender, all apelike proportions shed. Aiello and Schmid conclude that the newly evolved, slender build was an adaptation to greater activity, including the ability to be an effective runner. Walter Bortz, a physician in California, had published the same conclusion in 1985 in the *Journal of Human Evolution*, based on behavioral and physiological arguments.

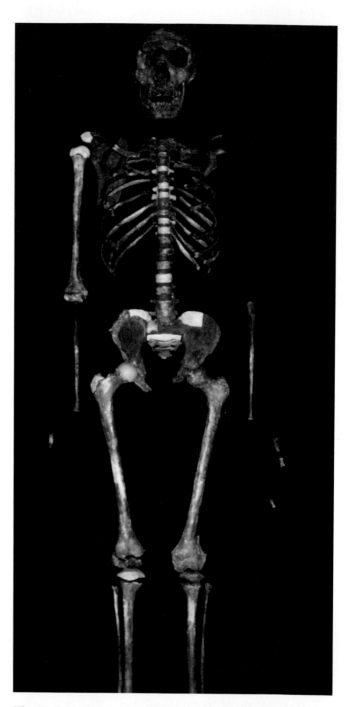

The Turkana boy, a 1.6-million-year-old Homo erectus *youth, was excavated in 1984 from the west side of Lake Turkana, Kenya.*

The second modification—concerning size—is as dramatic as this change in proportions and may have far-reaching implications. In all early hominids except *Homo*, estimates of average height fall in the region of 5 feet 3 inches. There was, however, a large difference in body size between males and females (sexual dimorphism). Australopithecine males could be as much as twice as heavy as females, a pattern common in primates and other mammals in which there is intense social competition among males for access to mates. In gorillas, for instance, mature males weigh an average of 350 pounds (females are less than 200 pounds), and such a male defends a "harem" of mature females against potential rivals.

When fossils of *Homo* appear in the record, average height increases by about 6 inches. Instead of being twice the bulk of females, moreover, as in *Australopithecus*, *Homo* males were only 20 percent bigger, a difference not much greater than is seen in modern humans. Such a change in sexual dimorphism can be interpreted as a shift from a social system characterized by intense sexual competition among unrelated males to one in which related males share in defending and having reproductive access to a group of females. Chimpanzees have such a system, and their sexual dimorphism is also about 20 percent. We may conclude that this new direction of hominid adaptation involved more than simply a shift in diet. Material technology became important for the first time, and possibly a social system emerged in which females left their natal group and male relatives remained, cooperating in their reproductive quest and, perhaps, in their search for food, as well.

How Many Species?

For about 1.5 million years the two types of hominid adaptation—small-brained/large cheek-toothed and large-brained/small cheek-toothed—coexisted. But a trend was developing: australopithecines were becoming extinct, the extreme version (*Australopithecus boisei*) persisting the longest. This trend may have been the result of resource competition from *Homo* species on one side and the burgeoning success of ground-dwelling baboons on the other. In any case, most anthropologists would argue that by about 1 million years ago, only one species of hominid existed in Africa: *Homo erectus*, the species that expanded into the lower latitudes of the Old World.

From half a million years ago onward there began to appear hominids that bore *Homo erectus*-like features together with certain modern ones, including expanded brain size. This new form of hominid, specimens of which have been found in Africa, Europe, and Asia, are generally known as "archaic *sapiens*," a term around which much disagreement and confusion swirl. Biologists like to classify organisms with the traditional combination of genus (like *Homo*) and species (like *sapiens*) name. Such a classification allows individual organisms to be recognized as members of a biological group within which members may breed. The use of the term archaic *sapiens* indicates an uncertainty on the part of anthropologists as to the status of such individuals: they are neither *Homo erectus* nor *Homo sapiens*, but may be a transitional form between the two. The term archaic *sapiens* is therefore more a descriptive shorthand than a formal classification. (The situation is further confused because Neanderthals, to which formal classifications have been applied, such as *Homo neanderthalensis* or *Homo sapiens neanderthalensis*, are also sometimes said to be examples of archaic *sapiens*.)

The history of the *Homo* lineage includes changes in anatomy, in technology, and in subsistence strategies, each of which contributes to an understanding and definition of the origin of modern humans. There must have been important interactions between these three aspects of our ancestors' history, but the links are not always as evident as might be expected. The general question at hand, however, concerns the evolutionary trajectory along which our ancestors eventually became modern: was it a steady, cumulative process or a relatively sudden, late transformation?

First, the anatomy. As always, the fossil record is patchy at best, and the record for early *Homo* is no exception. Although *Homo habilis* may have appeared first some 2.5 million years ago (based on fragmentary fossil evidence and the fact that stone-tool making began then), the earliest good fossil evidence is a little less than 2 million years old. The tall stature–low body bulk configuration so evident in *Homo erectus* almost certainly emerged with *habilis*. The transition from *habilis* to *erectus* involved a slight enlargement of the brain (to almost 900 ccs) and a change in facial features, including the development of prominent ridges about the eyes, the browridges. Once this configuration evolved, it persisted throughout the history of the *Homo* lineage— clearly a stable adaptation.

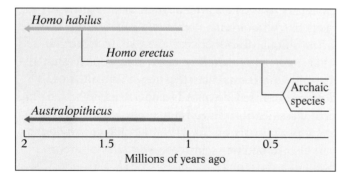

Although Homo erectus *roughly followed* Homo habilis *and was itself followed by the archaic* sapiens, *the pattern of change is uncertain. Rather than the simple unilinear progression depicted here, the pattern may have had branches representing several different* Homo erectus-*like or* Homo sapiens-*like species.*

With the origin of modern humans, however, there was a change in skeletal build—not so much in shape as in strength. *Homo erectus* bones are very much the same shape and size as those of *Homo sapiens*, but are more thickly buttressed—particularly the long bones. A cross-sectional view of the thigh bone, for example, shows the dense cortex in *Homo erectus* to be twice as thick as in *Homo sapiens*. This extra thickness of bone suggests the regular application of greater forces that would be experienced in an energetic daily life. The presence of big muscle attachments on *Homo erectus* bones is consistent with this picture. The origin of modern humans therefore involved a reduction in the physical force routinely experienced by individuals. In other words, modern humans were physically active, like *Homo erectus* individuals, but not as strong.

The lack of change in overall anatomy of *Homo erectus* during its 1.2 million years of existence contrasts with changes in the head. The earliest known *Homo erectus* crania, from East Africa, had brains in the region of 800 to 900 ccs in capacity; the face was much flatter than in the australopithecines, the cranium rounder and longer, and browridges developed over the eye orbits. The principal change in head anatomy by the time the threshold of *Homo sapiens* had been reached was in the size of the brain, now in the region of 1100 ccs. Because of the sparse fossil record, it is not possible to be certain whether this increase proceeded gradually throughout the period, or in one or more sudden increases.

To evolutionary biologists, the trajectory of evolutionary change—gradual versus punctuational—has been the subject of lively debate. Those who favor gradual change see evolution as a continuous response to selection pressure. Those who emphasize punctuational change interspersed between periods of stasis argue that factors relating to embryological development and population structure may constrain evolutionary responses to selection pressure; change, when it comes, does so rapidly. This latter pattern, jointly proposed in 1972 by Niles Eldredge, of the American Museum of Natural History, and Stephen Jay Gould, of Harvard, is known as punctuated equilibrium. Two decades of debate and empirical study indicate that both punctuational and gradual modes of evolutionary change occur, though there are still differences of opinion over their relative importance. For a decade Milford Wolpoff, of the University of Michigan, has been a leading spokesman in favor of a gradual mode of evolution throughout the history of *Homo erectus*, while Philip Rightmire, of the State University of New York, Binghamton, argues for a punctuational pattern. Their disagreement is indicative of inadequate data.

This question relates to another important discussion in the anthropological community, one that bears strongly on the origin of modern humans. The issue concerns the pattern of change between earliest *Homo* and *Homo sapiens*. Was it a simple, unilinear progression: *Homo habilis* to *Homo erectus* to *Homo sapiens*? Or did the pattern branch, with several *Homo erectus*-like and then *Homo sapiens*-like species evolving in different parts of the Old World through time, all but one eventually becoming extinct? Some scholars argue that what is called *Homo erectus* in Africa is different from *Homo erectus* in Asia, and that separate species names ought to be assigned to reflect this complexity. Others suggest further that the great anatomical variability seen among different geographical samples of so-called archaic *sapiens* (from 0.5 to 0.1 million years ago) indicates the existence of different species.

A straight-line evolution, or a branching pattern? Those who emphasize ecological factors support the probability of a branching pattern: an animal as widespread as *Homo erectus* and archaic *sapiens* would almost certainly differentiate into different geographic species, if the pattern of other mammals is any guide. Those who emphasize the unifying power of culture argue, conversely, that coexistence of more than one species at any given time is highly unlikely, giving a unilinear pattern. Even if the fossil record were much better, potentially allowing the question to be solved, different theoretical approaches still leave open the possibility of disagreement. The problem of what constitutes anatomical variability within or among species is faced by all paleontologists, who lack living populations with which to resolve it.

Technological Innovation

Whatever the true phylogenetic pattern in time, it is clear that over 1.2 million years hominid brain size—and presumably intellectual capacity—increased substantially, from almost 900 ccs to 1100, a 30 percent rise in a creature of the same body size. Technology, language, and social complexity are likely to have contributed to, or to have been enhanced by, this increase.

The idea of "Man the Toolmaker"—the concept that the manufacture and use of stone tools shaped our evolutionary career—was once extremely popular; its emphasis on technology still strikes a familiar chord in modern society. However, the products of stone-tool making—the artifacts themselves—do not evince dramatic progress or improvement over the 1.2-million-year period that we have been examining. Stasis in the number and type of tools produced is a much more apt description. The earliest stone tools in the record, from 2.5 million to 1.7 million years ago (the so-called Oldowan technology) were very basic. A stone pebble (the core), usually of lava, was struck using a second pebble (the hammer) to produce a crude chopper or scraper, plus many sharp flakes. Both the pebble tools and the flakes were probably utilized in various ways. There is little sense of systematic toolmaking in the Oldowan technology, more of opportunistic stone shaping or breaking with quick blows (knapping). The name of the technology derives from Olduvai Gorge in Tanzania, where Mary Leakey spent many decades (beginning in the 1930s) excavating and then describing the various tool types.

Coincident with the origin of *Homo erectus* 1.7 million years ago, a new technology developed, the so-called Acheulian technology. The principal difference between the Oldowan and the Acheulian was the addition of several larger implements, including the hand ax, the cleaver, and the pick. The hallmark of the Acheulian is perhaps the hand ax, a teardrop-shaped tool that required much more extensive work to produce than did any piece in Oldowan assemblages. The term Acheulian derives from the site of St. Acheul in France, where the first examples of hand axes were discovered in the nineteenth century.

The number of identifiable tool types in Acheulian assemblages is small: perhaps a dozen distinct forms, not all of which occurred at every site.

Not until archaic *sapiens* had evolved did this number increase substantially—a remarkable degree of technological stasis in human terms. Although the most skillfully made hand axes are found relatively late in the archeological record, there is no apparent steady improvement in skill through time: some of the later assemblages are as crude as some of those to be found at the beginning of the record.

As *Homo erectus* expanded from Africa into the Old World, tool technology naturally went with them. An interesting pattern developed, however, one that still requires complete explanation. Although typical Acheulian assemblages have been found throughout Europe and western Asia, none occur in East Asia. Instead, a chopping tool assemblage (like the Oldowan in some ways) appears in this part of the world. This pattern has been interpreted by some to indicate the existence of two different types—species—of people in the West and the East. Others reject this notion, suggesting that in the East bamboo

Representative samples of early stone tools of the Oldowan (left) and Acheulian (right) type. Oldowan tools are made from lava pebbles, and typically are choppers and scrapers, and many small, sharp flakes. Note the teardrop-shaped hand axe that characterizes Acheulian assemblages.

tools served the cutting and slicing functions that hand axes were used for in the West.

When archaic *sapiens* arose half a million years ago, complexity and sophistication of stone-tool assemblages began, as shown by innovation in the preparation of cores and the retouching of flakes. The number of tool types increased dramatically, as did the standardization of their manufacture.

On the face of it, therefore, the notion that the rise of technology provided the selection pressure for brain-size increase in *Homo erectus* is not supported. "It seems to me to be an inadequate explanation, not least because tool making can be accomplished with very little brain tissue," says Harry Jerison, a neurobiologist at the University of California at Los Angeles. Jerison bases his conclusion on the size of the brain areas that drive the motor activity involved in toolmaking. By contrast, he continues, "the production of simple, useful speech . . . requires a substantial amount of brain tissue." Perhaps, then, the evolution of language, not the development of technology, provided the evolutionary foundation upon which natural selection built the bigger brain?

Unfortunately, inferring language abilities in fossil humans is one of the least tractable problems that face anthropologists. Brain size itself is no certain signal. Tangible but indirect indications of language, such as symbolic image-making, are found only late in the archeological record. Does this imply that human species prior to modern humans were without spoken language?

Not necessarily. As we shall see, many scholars believe that spoken language reaches back to the beginning of the *Homo* lineage—indeed, that it was a prime cause of brain expansion. This position is that of gradualistic evolution of language capacities, and it contrasts with the opinion of other scholars, who argue that a substantial enhancement in spoken language capabilities was an important element in the biological shift to modern humans, perhaps even the sole element. This would be a punctuational pattern.

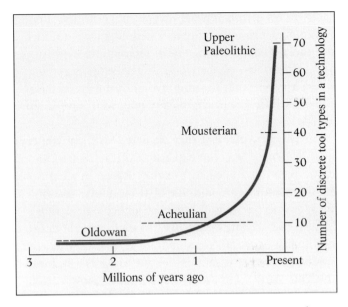

A trajectory of change in stone-tool technologies reveals that stasis typifies much of the prehistory tool record, until relatively recent times.

Debates over Hunting

Another aspect of behavior that impinges on potential expansion of intellectual powers in the *Homo* lineage is social complexity, an area encompassing activities related both to reproduction and to subsistence. There is no question that social interaction among higher primates is an exceedingly demanding exercise, requiring knowledge of the status of other individuals and of the alliances among them. For this reason primatologists have in recent years come to acknowledge that the skills required for successful social interaction have almost certainly contributed to the brain expansion that occurred during primate evolution. For instance, in a recent major assessment of cognition in nonhuman primates, Dorothy Cheney and Robert Seyfarth of the University of Pennsylvania and Barbara Smuts of the University of Michigan concluded: "The research we review suggests that among nonhuman primates, sophisticated cognitive abilities are most evident during social interactions with [other members of the troop]." The same argument may carry through to brain evolution in the human family, where social complexity may well have intensified.

The new element that occurred with the appearance of *Homo* was the expansion of the diet to include meat, as indicated to some degree by dentition, by the manufacture and use of stone tools, and by the presence of animal bones at archeological sites. Very probably this broadening of diet contributed to the ability of *Homo* eventually to expand beyond the traditional range of hominids, out of Africa and into the Old World. For *Homo sapiens*, the combination of hunting meat and gathering plant foods became the primary subsistence adaptation, one that began to be replaced by 10,000 years ago with the gradual adoption of food production. Until recently anthropologists were still able to study the lifeways of such foraging people, small populations of which persisted in Africa, Asia, and North and South America. Despite very different environments, foraging lifeways had certain features in common. For instance, people often lived in small, mobile bands of about 25 individuals, which were members of a larger group, or tribe, of some 500 individuals. Economic activity was often divided, with males principally responsible for hunting and females the principal gatherers of plant foods (typically, the bulk of the diet). The hunting and gathering lifeway, given its broad geographical distribution, was clearly a successful adaptation.

How soon in the history of the *Homo* lineage hunting and gathering evolved is crucial to an understanding of the eventual transition to modern humans. Archeologists have long known that stone tools and animal bones have been found in association from the earliest part of the prehistoric record, close to 2 million years ago. With the model of primitive foraging societies in mind, scholars interpreted these associations as an indication that early *Homo* had begun to establish at least the rudiments of the hunting and gathering lifeway. Co-occurrences of bones and stones were taken to be the remnants of ancient campsites: the temporary living sites of a mobile, foraging people. With this went the notion, again based on modern experience, that a division of labor had developed. In other words, archeologists assumed that hunting and gathering has a long history, perhaps with a steadily increasing sophistication toward that of *Homo sapiens*. About 15 years ago, Glynn Isaac and Lewis Binford independently initiated investigations to test the validity of these assumptions.

Binford, an archeologist at Southern Methodist University, Dallas, examined bone assemblages that had been found at archeological sites. He compared them with the composition of remains found in the dens of, for example, hyenas, a scavenging species, and concluded that there was little difference between them. He quickly developed the opinion that archeological evidence from the early part of the human record had been grossly overinterpreted, that the bone accumulations associated with stone tools were nothing more than the remains of occasional scavenging events. The co-occurrences of bones and stones were not indicative of campsites in any sense. The hunter-gatherer way of life, Binford concluded, was a very recent development. "Between 100,000 and 35,000 years ago the faint glimmerings of a hunting way of life appeared," he wrote. "Our species had arrived—not as a result of gradual, progressive processes but explosively in a relatively short period of time." The immediate ancestors of modern humans were little more than "marginal scavengers," he asserted. In Binford's assessment, the evolution of human foraging had followed a punctuational pattern, with a recent shift to the modern behavior.

Isaac, an archeologist at Harvard University before his death in 1985, came to a different conclusion. He initiated new excavations designed specifically to test the hypothesis that the co-occurrence of bones and stones was the result of direct hominid activity. With a large team of colleagues, he exposed a 1.5-million-year-old site east of Lake Turkana in Kenya, and he demonstrated that the remnants of bone had been brought to that spot, that stone tools had been made there, and that some of the tools had been used to remove meat from the bones. Microscopic analysis of the edges of a selection of stone flakes revealed that some had also been used on wood and some on soft plant material.

These various lines of evidence suggest that the site, which had been on the bank of a small river, was the focus of varied subsistence activity, not mere scavenging at a carnivore kill site. Isaac remained cautious, however, and said that it was impossible to prove that the location had been a campsite in the sense understood for modern foraging people. "I believe there was a complexity to subsistence in early hominid life," he finally concluded. "Hunting and gathering developed slowly in human history, and its rudiments were most definitely present before the origin of modern humans." Isaac's view, which probably represents the current majority opinion among archeologists, was of a gradual evolution of the foraging adaptation that came to characterize modern human subsistence prior to agriculture.

An Evolutionary Bridge

The long prelude to modern humans, therefore, was not a preparation for a final event, simply its history. With a knowledge of that history, a clearer sense of *Homo sapiens* in relation to Huxley's "gulf between . . . man and the brutes" becomes possible. With an appreciation of the kinds of creatures that populated our history, that gulf is bridged.

2

PRELUDE TO THE MODERN DEBATE

Joachim Neander, poet and composer, died in 1680. Toward the end of his short life he frequently visited a quiet valley near his home to seek his idyllic muse. As a mark of respect for this young man, the local burghers named the valley after him: Neander Thal. An accident of history ensured that this name would gain more than topographical fame. In August 1856, with the Prussian construction industry hungry for raw material, quarry workers in the Neander Valley unearthed human bones in a cave, Feldhofer Grotto, above the Düssel River, flowing toward its confluence with the Rhine at Düsseldorf. Probably an entire skeleton had been entombed in the limestone sediments of that cave; all that survived the quarrymen's zeal, however, was the top of a cranium, some leg and arm bones, and other damaged parts.

Representations of Neanderthals have often erroneously emphasized a putative brutishness and dim-wittedness.

The fossilized bones were taken to Carl Fuhlrott, a mathematics teacher and local historian known to be interested in natural curiosities. Fuhlrott recognized that the bones had certain unusual characteristics, particularly their thickness and heavy build compared with those of modern humans. Clearly the remains of a bulky and powerfully muscled individual, they were unlike anything Fuhlrott had seen before, so he sought the more informed viewpoint of Hermann Schaaffhausen, professor of anatomy at the University of Bonn. Schaaffhausen, agreeing with Fuhlrott that the bones belonged to a primitive form of human, described them with that interpretation at a gathering of the Lower Rhine Medical and Natural History Society in Bonn on 4 February 1857, almost three years before *The Origin of Species*. Thereafter, Schaaffhausen and Fuhlrott publicly presented this view many times, sometimes independently, sometimes jointly.

According to Schaaffhausen, the unusual robusticity and form of the bones were natural aspects of the anatomy. The individual had probably belonged to "one of the wild races of northwestern Europe," people antecedent to the Celts and the Germans. "It was beyond doubt that these human relics were traceable to a period at which the latest animals of the Diluvium existed," he said in 1859, the year of publication of Charles Darwin's *Origin*. (A term of some imprecision, no longer in use, the Diluvium referred to what were considered to be the debris of the Noachim flood. In reality, the deposits of the Diluvium mark the boundary between glacial and postglacial times, or the Pleistocene and the Recent.) In the view of Professor Schaaffhausen, therefore, the fossil man from the Neander Valley was from a distant part of human history—an ancestor of sorts.

Thus began a long and continuing assessment of Neanderthal Man's place in human evolution, a century and a quarter of ambivalence on the part of the anthropological profession. As we saw in the Prologue, the issue is this: Was Neanderthal Man a direct ancestor of modern man? Or was he a side branch, leaving no descendants among modern populations? Professional opinion, which has flipped back and forth several times, remains unresolved in the broader context of today's debate on the origin of modern humans.

Before we go on to explore this current debate, with its unexpected sources of new data, a brief look at its antecedents will be enlightening. Science progresses by constantly testing hypotheses; established ones are sometimes eliminated and new ones substituted. There is reasonable expectation, therefore, that today's discussion is closer to what actually happened in human prehistory than earlier explanations. Nevertheless, an appreciation of the history of a scientific question—especially one in a historical science like anthropology—can reveal how different scholars may place different interpretations upon the same evidence because of their differing theoretical preconceptions.

Neanderthal Is Confined to Obscurity

Two key questions were immediately raised over the identity of the Feldhofer Neanderthal: How old were the fossils? How did the unusual anatomy relate to that of other humans? The intellectual context in which they were asked was far from stable.

Ideas on evolution and geology were, at this period, closely linked together. For many years Earth history had been interpreted in terms of catastrophism, the notion that the succession of epochs visible in the geological record reflected episodic violent revolutions caused by divine intervention. With each catastrophe all existing life forms were destroyed, new ones then being created to restock the Earth. Baron Georges Cuvier (1769–1832), a notable French paleontologist, was the author of the theory, behind which the scientists of continental Europe stood solidly. In Britain, meanwhile, James Hutton (1726–1797) and then Sir Charles Lyell (1797–1875) were the authors of the uniformitarian view that geological change was gradual, not episodic and violent. Catastrophism was necessarily anti-evolution; uniformitarianism, which was gaining stronger and stronger support through the middle of the nineteenth century, eventually came to be a partner of the evolutionary viewpoint. Not, however, at first.

Lyell, who never fully embraced his friend Charles Darwin's evolutionary theory, had an opportunity to see the Feldhofer Neanderthal fossils not long after they were discovered. To him, as to Schaaffhausen, the bones looked ancient, but he argued that given the absence of animal bones from the same sediments, it was impossible to ascertain their age. For geologists of Lyell's time, who had no physical means of dating rock sediments, the evolutionary stage of animal fossils in a deposit was a guide to the age of the deposit—a technique known as faunal correlation. Animals at an early stage of an evolutionary sequence are found in rocks of an earlier age than those later in the sequence, for instance. Modern geologists still use this technique, but also enjoy a range of physical methods for dating sediments, often based on radioisotopes. Alas, continued quarrying activity at Feldhofer obliterated any reasonable hope of using modern methods to obtain a reliable age for the Neanderthal fossils.

Thomas Henry Huxley, an enthusiastic supporter of evolution who examined casts of the fossil bones, concluded that they represented merely an extreme form of historical human being, "like certain ancient people who inhabited Denmark during the 'stone period.'" Huxley was eager to find evolutionary links between ancient apes and modern humans, but Neanderthal Man, primitive though he was in many ways, did not fit the bill. Here Huxley was facing paleontology's perennial puzzle: What degree of anatomical difference constitutes real biological difference? Huxley, familiar with modern human anatomy, could see that Neanderthal anatomy was different in many ways, but he concluded that it represented merely a primitive version of modern anatomy, not a distinct earlier form.

With no known human fossils available to them, anthropologists of this era were able to erect only the vaguest of human evolutionary trees. Huxley and others postulated the existence of an apelike ancestor transformed over an unknown period of time and through less and less primitive stages into the modern human form. Huxley was particularly impressed by the size of the Neanderthal brain, which was at least equal to that of modern humans. Creatures with such brains were already essentially human, he concluded.

Some scholars were thus prepared to view the Neanderthal bones in an evolutionary context and, to differing degrees, were successful in so doing. But many dismissed the idea entirely. The bones were modern, they asserted, the remains of a crazed or diseased individual. One proposed explanation was that a Russian Cossack pursuing Napoleon west in 1814 had become ill and crawled into the cave to die; the "Cossack's" bowed legs, it was noted, were clearly the result of life on horseback. Another hypothesis came from the pen of an Englishman, J. W. Dawson: "It may have been one of those wild men, half-crazed, half-idiotic, cruel and strong, who are always more or less to be found living on the outskirts of barbarous tribes, and who now and then appear in civilized communities to be consigned perhaps to the penitentiary or the gallows, when their murderous propensities manifest themselves." Still another assertion was that the Neanderthal's extraordinary anatomy, including the bowed legs, was the result of childhood rickets, the pain of which had caused the unfortunate individual constantly to furrow his forehead—hence the prominent browridges.

All such speculation was put to an end in the 1870s when Rudolf Virchow, the prominent German anatomist and pathologist, pronounced the bones to be modern and pathological, not ancient. The renowned pathologist, distinctly unsympathetic to the theory of evolution, agreed with the suggestion that the Feldhofer individual had suffered from rickets. But he had been old when he died, Virchow said,

Changing Views of Earth History

We are used to thinking about Earth history in terms of the inexorable accumulation of small changes over long periods of time. Dramatic events such as volcanic eruptions, earthquakes, and tidal waves occasionally bring instant, local transformation; but the configuration of the continents, the building of mountains, the layering of massive sediments, and erosion by the elements each proceed at imperceptible rates to, in combination, shape the major features of the Earth. We understand that slow, incremental geological change forms the context in which the gradual process of evolution unfolds. Earlier views of Earth history were very different.

Three and a half centuries ago Archbishop James Ussher (1581–1656) announced that his biblical calculations showed Earth history to have begun in 4004 B.C.E. Some 6000 years, therefore, comprised the Creation, the Noachim Flood, and the subsequent history of the human species. This Western intellectual view of the world was essentially theological. Until the late eighteenth century the notion that a species might become extinct was regarded as an affront to the guardianship that the Creator maintained over His creations. Life on Earth was perceived in the context of what was known as the Great Chain of Being, an ordering of organisms from the lowliest forms to the most perfect—humans were a little lower than the angels. The chain was meant as a literal description of organisms as God had created them, and its continued integrity was fundamental: lose a link in the chain and the whole would be threatened. Fossilized bones, once explained as sports of nature, were eventually encompassed into this worldview as relics of the casualties of the Noachim Flood, the Deluge described in Genesis. The deposits that entombed the fossils were known as the diluvium, and the governing theory the Diluvial theory.

Paradoxically, clergymen, as avid amateur geologists, played a major part in changing this view, as did the gathering pace of the Industrial Revolution: in the late eighteenth century the digging of canals and the excavation of coal mines exposed successive sedimentary layers of rock and the fossils they contained. Increasingly it became apparent that some of the fossilized bones belonged to species that no longer existed; in some cases a progression of forms could be traced from lower to higher sedimentary layers. Since Noah was supposed to have rescued all living species, these discoveries revealed that the biblical account must be incomplete.

In 1808 Baron Georges Cuvier (1769–1832), a zoologist and paleontologist at the Paris Natural History Museum, suggested that there had been two great deluges, not one; the second was the Noachim Flood. This theory, which came to be known as catastrophism, was warmly embraced by intellectuals in Europe, since it accepted scientific observation while maintaining much of the biblical account. As knowledge of the geological and fossil records accumulated, further abrupt changes were noted, with the loss of certain species—eventually rationalized as the result of a series of deluges, not two. Adam Sedwick (1785–1873), a mentor of Charles Darwin at Cambridge University, was a great proponent of catastrophism, and described Earth history as follows: "At succeeding periods new tribes of beings were called into existence, not merely as the progeny of those that had appeared before them, but as new and living proofs of creative interference." Instead of a single Creation, as described in the Old Testament, Earth history was to be seen as a series of creations. This intellectual outlook was, in its way, as theologically based as before.

Such a hectic history of life on Earth, with its dozen or so successive catastrophes, would have ill-fitted the time allotted according to the Ussher calendar. Fortunately for proponents of catastrophism, Comte Georges de Buffon, keeper of the Jardin de Roi in Paris, produced a revised age of the Earth based on calculations for the rate of cooling from

A view of the biblical deluge, Luther Bible, 1534. The theory of catastrophism envisaged a series of world-engulfing catastrophes, which wiped out all forms of life, followed by a restocking by new, more advanced forms.

a molten state: 74,832 years, with life appearing a little over halfway through.

The theory of catastrophism soon found itself in competition with a new hypothesis: uniformitarianism, which views the major geological features of the Earth as the outcome of everyday, gradual processes, not occasional violent events. James Hutton (1726–1797), a Scot, seeded the ideas of uniformitarianism, but it was Charles Lyell (1797–1875), also from Scotland, who solidified the ideas, effectively becoming the founder of modern geology. Both men, impressed by the power of erosion they observed, reasoned that with sufficient time major geological features could be fashioned by such forces. According to proponents of uniformitarianism, the abrupt changes seen in the geological record were artifacts of incomplete data; further research one day would fill in these gaps.

Lyell published *The Principles of Geology* in three volumes, the first of which appeared in 1830. The conclusion of uniformitarianism—specifically, that the Earth is unimaginably old—was important for Charles Darwin's development of the theory of natural selection, based on the accumulation of small changes over long periods of time. Like Lyell, Darwin believed that the abrupt changes seen in the prehistoric record were a consequence of its incompleteness, not a real phenomenon of nature. Neither Lyell nor Darwin treated seriously the notion of occasional mass extinctions—it must have seemed to them too much like catastrophism. As a result, mass extinctions have become the subject of serious geological research only during the past several decades.

Five major extinction events, in fact, appear in the record, during which as many as 95 percent of existing species became extinct. At least one such event, at the boundary between the Cretaceous and Tertiary, 65 million years ago, was almost certainly caused by Earth's impact with an asteroid or comet. Some researchers, in particular David Raup and Jack Sepkoski of the University of Chicago, suggest not only that the other four major events are the result of asteroid impact, but also that similar but smaller impacts have occurred every 30 million years throughout Earth history. Ultimately, suggests Raup, 60 percent of species extinctions result from asteroid impact. If he is correct, modern geology may become to some extent a New Catastrophism, without the theological content.

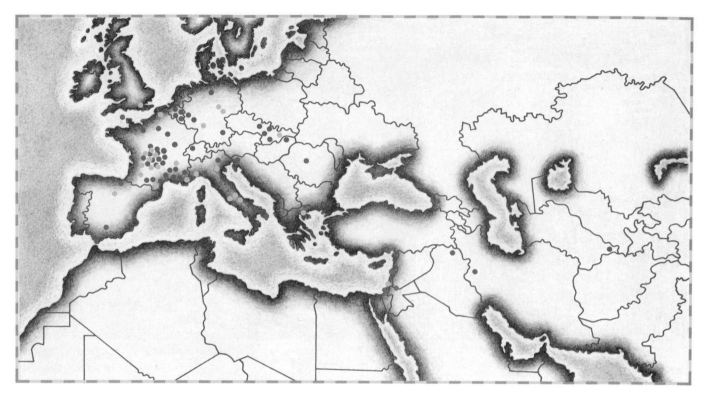

Distribution of Neanderthal fossil sites: The Neanderthal range appears to have been restricted to Europe, the Near East, and Central Asia.

and since primitive foraging societies simply could not support people to such an advanced age, he must have lived in a sedentary—and therefore recent—society. The bowed shape of the Neanderthal thigh bones is indeed reminiscent of what occurs in rickets, but rachitic bones are thin and porous as a result of the loss of calcium, not thick and robust as in the Feldhofer Neanderthal. Virchow's misinterpretation of the evidence almost certainly was the result of his disinclination to accept any kind of evolutionary scenario, which was becoming more and more popular in Europe through the influence of Charles Darwin in England and Ernst Haeckel (a student of Virchow's)

in Germany. Nevertheless, Virchow's scholarly stature was such that his assessment of Neanderthal prevailed, at least for a while.

"Of course this is amusing now," notes William W. Howells, an emeritus professor of anthropology at Harvard and an expert on the fate of the Neanderthals, "but we should be fair and remember that these anatomists were among the best men of their day. If you have the skeleton of that first Neanderthal before you, and at the same time try to imagine yourself a hundred years back, it is easier to comprehend the disagreements." The incomplete specimen had no face and was of uncertain geological age: in any case, at first there was no strongly developed evolutionary ethos within which to interpret it.

Initially accepted into the folds of human history by Hermann Schaaffhausen, Neanderthal Man was subsequently consigned to anthropological obscurity. The first flip was complete. By the end of the century, however, the discovery of more and more fossil individuals with the same suite of curious anatomical characteristics effectively undermined the notion that pathology explained the appearance of the

Feldhofer individual. Subsequent discoveries of Neanderthal bones show that such beings lived between about 150,000 and 34,000 years ago throughout much of Europe and into the Near East; whatever their role in human prehistory, their population was by no means insignificant. More important, in the early 1890s the Dutch scientist Eugene Dubois discovered even more primitive-looking fossils, the so-called *Pithecanthropus erectus* remains, in Java, Indonesia. With a smaller brain than Neanderthal but the same robusticity of skeletal structure, these remains eventually came to be known as *Homo erectus*.

Eugene Dubois was convinced he would find early human fossils in Java, and did in the 1890s. He called his finds Pithecanthropus erectus, *now known as* Homo erectus.

A Brief Reprise

Those who had been looking in vain for a direct role for Neanderthal in human evolution often lamented his lack of sufficiently apelike characteristics. With the *Pithecanthropus* fossils to hand, a solution was articulated at the turn of the century by Gustav Schwalbe, a professor at the University of Strasbourg. Schwalbe, a supporter of Darwinian theory, spent several years reexamining the *Pithecanthropus* and Neanderthal remains in the context of possible evolutionary relationships. With its smaller brain and pronounced browridges, *Pithecanthropus* was clearly more primitive—that is, apelike—than modern humans, observed Schwalbe, while the Neanderthals, less primitive than *Pithecanthropus*, were nevertheless distinct from modern humans. Where Huxley had considered the anatomical differences between Neanderthals and modern humans as not representing significant biological difference, Schwalbe concluded otherwise. He therefore suggested that both *Pithecanthropus* and Neanderthal were part of a steady progression from primitive to modern human beings. Arguing for a straight line of change through time,

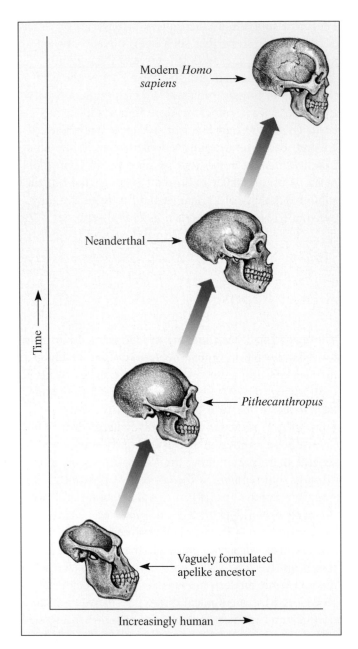

Modern *Homo sapiens*

Neanderthal

Time →

Pithecanthropus

Vaguely formulated apelike ancestor

Increasingly human →

At the turn of the century, Gustav Schwalbe developed the unilinear hypothesis, which described the evolution of modern humans from an apelike ancestor through Pithecanthropus *and Neanderthal intermediates. Hrdlicka also supported this model.*

the pattern known as unilinear evolution, Schwalbe thereby aligned himself with Schaaffhausen's original view of Neanderthal.

Schwalbe's proposal was welcomed by some scholars in continental Europe and, much more enthusiastically, among British anthropologists. But this particular rebirth of the idea of a unilinear progression with no side branches, no bushiness to the human evolutionary tree, was to be short-lived. Shortly before the First World War, the combination of two separate but related events effectively expelled Neanderthal from human ancestry, consigning it this time not to anthropological obscurity (as a pathological individual) but to evolutionary oblivion (as an extinct side branch of the human tree). These events were the discovery in 1908 (and subsequent misinterpretation) of a famous Neanderthal skeleton at the site of La Chapelle-aux-Saints; and the discovery in 1912 (and subsequent misinterpretation) of parts of a fossil skull that came to be known as Piltdown Man.

An Intellectual Agenda Is Set

The importance of the La Chapelle-aux-Saints Neanderthal, unearthed in a cave near a village of that name in the Dordogne region of France, lay in its completeness. For the first time, anthropologists had an opportunity to compare in detail Neanderthal anatomy with that of modern humans and with other forms considered to be ancestral to modern humans.

Sir Arthur Keith (1866–1955), one of the most prominent British anthropologists in the first decades of this century, long held the conviction that the modern human form was very ancient. His view was that so specialized an animal as *Homo sapiens*, particularly so elevated a structure as the human brain, demanded a long time in which to evolve. Neverthe-

The Dordogne Valley in southwest France has long been the focus of archeological studies, with many sites of Neanderthals and modern humans and, of course, many decorated caves.

less, in common with several of his eminent colleagues, Keith briefly espoused Schwalbe's unilinear hypothesis. Keith's view was eventually swayed by the evidence of two skeletons: Galley Hill Man, unearthed east of London in 1888 but not properly studied until 1910; and Ipswich Man, discovered in 1911 near the town of Ipswich, in the east of England.

Both skeletons were undoubtedly modern and yet, to judge from the age of the sediments from which they were recovered, appeared to be ancient, as ancient as the Neanderthals. (As it later turned out, the two individuals had been buried at death, so their bones came to lie in deep, ancient sediments—a phenomenon known as intrusive burial. Keith and others simply made the mistake of assuming that the

The Neanderthal skeleton found in 1908 at La Chapelle-aux-Saints, France, played an important role in the developing views of the humanness of the species.

thal anatomy had been exaggerated. This left viable the possibility that Neanderthals were part of an evolutionary progression to modern humans that included forms such as Ipswich and Galley Hill Man. In other words, Schwalbe's unilinear scheme might remain intact and Keith's rejection of it be an error. The argument could be settled only by the discovery of a complete Neanderthal skeleton.

"It was just such a skeleton that was unearthed in August 1908 by three clerical historians, the Abbés J. and F. Bouyssonie and L. Bardon," wrote Michael Hammond, a historian of science at the University of Toronto, referring to the discovery of the La Chapelle-aux-Saints remains. The subsequent analysis of this extraordinary skeleton by the renowned paleontologist Marcellin Boule was to have a profound impact on the intellectual agenda of anthropology for the next four decades.

The Caveman Caricature

The La Chapelle-aux-Saints skeleton was sent, in early November 1908, to the Museum of Natural History in Paris, where Boule reigned as a towering intellect. The École d'Anthropologie might have been a more natural choice, but the Abbés Bouyssonie and Bardon were naturally antagonistic to the anticlerical tradition of the school. They sought the advice of that other giant figure of French prehistorical scholarship, the Abbé Henri Breuil, who assured his clerical colleagues that the skeleton would be in good hands with his friend Boule—a piece of advice that may well have changed the course of anthropological history.

Within a month of receiving the skeleton Boule made his first public observations, at a gathering of the Academy of Sciences. He estimated the age of the individual at death as between 50 and 55 years, so

skeletons had lived in an era as old as the sediments in which their bones were found.) Initially, however, these modern individuals and the more primitive Neanderthal were thought to have coexisted—in which case, Keith reasoned, Neanderthals could not be ancestral to modern humans and must be an extinct side branch.

The fragmentary nature of the Neanderthal fossils then known, however, allowed the counterargument that the putatively primitive aspects of Neander-

The French paleontologist Marcellin Boule was highly influential in the early decades of this century and championed the view that Neanderthals were not ancestral to modern humans.

the skeleton quickly became known as the Old Man of La Chapelle-aux-Saints. "[The cranium] strikes one first by its very considerable dimensions, keeping in mind the small stature of its ancient possessor. Next it strikes us with its bestial appearance, or, to put it better, by the general collection of simian or pithecoid characteristics." The term pithecoid (from the Greek word for ape) was often used to mean primitive, even if apelike characteristics were not strictly present; in this case, the few anatomical features that might truly be said to resemble those of an ape were the prominent browridges, the protruding face, and the long, low skull with receding forehead. Boule delivered his anatomical assessment: "It seems to me [that] . . . the group of Neanderthal-Spy-La-Chapelle-aux-Saints represents an inferior type closer to the Apes than to any other human group. Morpho-logically it appears to be placed exactly between the *Pithecanthropus* of Java and the more primitive living races, which, I hasten to say, does not imply, in my opinion, the existence of direct genetic descent."

Over the following four years Boule extended his analysis, producing an extensively detailed description and interpretation that he presented from time to time to gatherings of the academy and in a series of scientific papers. The picture Boule sketched of the Old Man of La Chapelle—and, by implication, all Neanderthals—was effectively that of a slouching, bent-kneed, bent-hipped semi-idiot, a characterization with which many of the most prominent British anthropologists, including Arthur Keith and Sir Grafton Elliot Smith (1871–1937), quickly concurred.

"His short, thick-set, and coarsely built body was carried in a half-stooping slouch upon short,

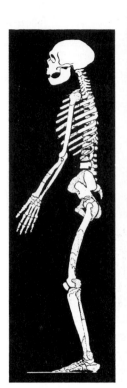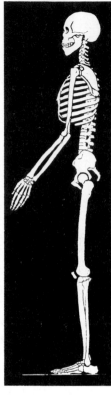

Marcellin Boule's examination of the La Chapelle-aux-Saints skeleton led him to regard Neanderthals as stooping and bent-kneed—a caricature of the caveman. He illustrated his view by a sketch, published in his 1923 book Fossil Men.

powerful, and half-flexed legs of a peculiarly ungraceful form," wrote Sir Grafton Elliot Smith, in his *Essays on the Evolution of Man* (1924), drawing largely on Boule's published works. "His thick neck sloped forward from the broad shoulders to support the massive flattened head, which protruded forward, so as to form an unbroken curve of neck and back, in place of the alternation of curves which is one of the graces of the truly erect *Homo sapiens*. The heavy overhanging eyebrow-ridges and retreating forehead, the great coarse face with its large eye-sockets, broad

nose, and receding chin, combined to complete the picture of unattractiveness, which it is more probable than not was still further emphasized by a shaggy covering of hair over most of the body." No evidence as to how hairy Neanderthals were existed, of course; Elliot Smith was filling out his fantasy of an uncouth brute that "belongs to some other species than *Homo sapiens*." (As early as 1864 a distinct species name had been coined by William King, professor of geology at Queens College, Galway, Ireland: *Homo neanderthalensis*.)

As University of Michigan anthropologist Loring Brace observed in a classic paper published in 1964, "Boule . . . depicted the Neanderthals in terms which have served journalists and scholars ever since as the basis for the caricature of the cave man." Boule produced drawings of Neanderthal Man and comparisons with modern humans that can be seen as prototypes for comic strip portraits. "Since he was not prepared to accept such a creature in the human family tree, he settled the question to the general satisfaction by declaring that the Neanderthals . . . became extinct without issue," noted Brace. Boule's analysis endorsed a picture of human evolutionary history that was far from unilinear. Neanderthal was one—extinct—side branch.

A Consensus Is Reached

Boule and his supporters held clear views of what kind of creature Neanderthal was, although terms such as "brute," "uncouth," and "unattractive" now sound less than scientific. In his *Fossil Men*, published in 1923, Boule acknowledged that "the Man from La Chapelle had a brain as large as that of the most civilized of modern races." Nevertheless, when he compared the overall shape of the Neanderthal skull with those of a chimpanzee and a modern Frenchman, Boule inferred significant differences.

Specifically, he noted that in chimpanzees the facial region is almost as large in side view as the cerebral region, whereas in the Frenchman the facial region is relatively small. Neanderthals, wrote Boule, are intermediate between these two, and the Neanderthal cranium is long and low, like the chimpanzees'. Boule inferred from these general similarities in shape between the crania of chimpanzees and Neanderthals a similarity in cognitive abilities—or at least that Neanderthals did not match the mental capacities of the modern Frenchman: "Thus there disappears, or is greatly lessened, the paradox seemingly indicated by the magnitude of the absolute volume of the La Chapelle skull, when due account is taken of the numerous signs of its structural inferiority." Boule thereby convinced himself that "other things being equal, the brain is relatively less than the brains of modern large-headed men."

Elliot Smith, perhaps the most eminent neurologist of his time, agreed with Boule's assessment. "Many recent writers have been puzzled to account for the great size of his brain, seeing that the average capacity of the Neanderthal cranium exceeds that of modern Europeans," he wrote in *Essays on the Evolution of Man.* "But . . . the development of the brain of Neanderthal Man was partial and unequal. That part of the organ which plays the outstanding part in determining mental superiority was not only relatively, but actually, smaller than in *Homo sapiens.*" In other words, the frontal region of the brain was smaller than in modern humans, according to Elliot Smith (a contention that later studies have not confirmed).

Elliot Smith speculated that the Neanderthal's large brain "was due to a great development of that region which was probably concerned with the mere recording of the fruits of experience, rather than the acquisition of great skill in the use of the hand and the attainment of the sort of knowledge that comes from manual experiment." Neanderthals, he wrote, were "clearly on a lower plane" than modern humans, a fact that accounted for their "failure in the competition with the rest of mankind." Very quickly and with almost fervent enthusiasm, the anthropological establishment accepted Boule's characterization of Neanderthal Man and pronounced the species an evolutionary specialization that went nowhere.

Enter Piltdown Man

Boule's description of Neanderthal left modern man ancestorless. In 1912, one of the most famous "fossils" of all time—the Piltdown Man—appeared, filling this void in our evolutionary past.

Charles Dawson (1864–1916), an amateur archeologist, recovered several pieces of human skull from a gravel pit near Piltdown Common, in the south of England, between 1908 and 1911. In May 1912 he took the fragments to the British Museum in London for evaluation. Sir Arthur Smith Woodward (1864–1944) thought the fossils might indeed be important, and he joined Dawson in further excavations. Pierre Teilhard de Chardin (1881–1955), the French priest and anthropologist, also joined the search. More skull fragments were unearthed, and a fragment of jaw. Smith Woodward studied the fossils and presented his results at a meeting of the Geological Society, London, on 12 December 1912. He described the skull fragments as indicating a form of human with a large brain, modern in aspect but with some primitive features. The jaw, he said, was apelike—it would have projected, as in chimpanzees and gorillas—but had some advanced features. Together, the fossils bespoke a stage in human evolution prior to the modern form.

The combination of a large brain and an apelike jaw, recovered from ancient deposits, provided an ancestral form that fit prevailing theory. Smith Woodward spoke for much of the British anthropological community when he said the discovery "tends to sup-

Workers at the Piltdown site, Sussex, England (circa 1912–13). Left to right: Robert Kenward, Jr. (standing), Charles Dawson (sitting), "Venus" Hargreaves (center), Arthur Smith Woodward and the goose "Chipper."

port the theory that [Neanderthal Man] was a degenerate offshoot of early man, and probably became extinct; while surviving man may have arisen directly from the primitive sources of which the Piltdown skull provides the first discovered evidence."

The prevailing theory was assembled from several lines of argument. First was Boule's rejection of a unilinear pattern of human prehistory and his conclusion that Neanderthal was a specialized offshoot of

human prehistory. This left open the need for an early human form that possessed a range of primitive features, specifically those that might be found in apes. Second was Keith's conviction that the modern human form had an extremely ancient history, ultimately reaching back to an apelike ancestor. Third, Elliot Smith was convinced that in human evolution brain expansion was the prime mover; in 1924, for instance, he wrote, "the brain attained what may be

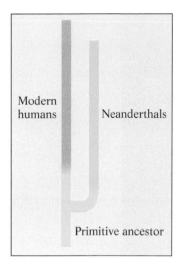

In the early years of the twentieth century, the presapiens hypothesis prevailed, supported prominently by British anthropologist Arthur Keith and French anthropologist Marcellin Boule; this stated that an early split occurred in human evolution, one branch leading to Neanderthals the other to modern humans.

termed the human rank at a time when the jaw and face, and no doubt the body also, still retained much of the uncouthness of Man's simian ancestors." Last, William Sollas, professor of geology at Oxford University, had developed the notion of mosaic evolution: that different parts of the body may change at different rates.

Although the British anthropological establishment—most prominent of whom were Keith, Elliot Smith, and Smith Woodward—enthusiastically welcomed the Piltdown material as tangible evidence that confirmed their theories, scholars in the United States and continental Europe were dubious, and critics questioned whether the fossils had come from two separate creatures, human and ape. Boule, who supported the overall evolutionary pattern promulgated by Keith, was one such critic. German anthropologists were uniformly skeptical, while in the United States Aleš Hrdlička, curator of physical anthropology at the National Museum of Natural History (Smithsonian Institution) and founder of the American Association of Physical Anthropologists, and

Henry Fairfield Osborn, director of the American Museum of Natural History, were supporters. Gerrit Miller, of the Smithsonian Institution in Washington, D.C., was a critic; in 1915 he said, prophetically, "Deliberate malice could hardly have been more successful than the hazards of deposition in so breaking the fossils as to give scope to individual judgement in fitting parts together." Since the jaw bone lacked the point of articulation with the skull, the nature of the fit between the two could not be checked.

The Second Flip Is Complete

Despite the differences of opinion over the interpretation of the actual fossils, Piltdown Man inexorably became part of the evidence supporting the new view of human prehistory. Boule's characterization of the La Chapelle-aux-Saints skeleton was the second major piece of evidence in favor of this new view, which was known as the presapiens theory. This theory held, in the words of historian of science Frank Spencer, that "there had been an ancient split in the human lineage which led to the early appearance of a relatively modern skeletal form alongside a more archaic hominid, represented in the fossil record by the Neanderthals." With the advent of this quintessentially branching view of human evolutionary history, the second flip was complete.

Deep Genetic Roots

From its inception in the second decade of the century, the presapiens theory effectively dominated anthropological thinking for almost 50 years, despite

German anatomist Franz Weidenreich, who began his career under Gustav Schwalbe at the University of Strasbourg in 1899, was an important figure in the development of studies of Pithecanthropus *and later of the* Sinanthropus *(also* Homo erectus*) fossils excavated from the Peking (Beijing) Man site in China.*

various vigorous efforts to dislodge it. Aleš Hrdlička attempted in the 1920s to resurrect the unilinear hypothesis, a mode of evolution that he had grown to support as a result of his study under Léonce Manouvrier (1850–1927) at the École d'Anthropologie, Paris, in the late 1890s. He rested his argument for what he called the "Neanderthal phase of Man" on two main points. First, contrary to popular scientific opinion, there was considerable anatomical variability among the available Neanderthal speci-

mens, which, he suggested, could be interpreted as evolutionary progression. Second, his examination of the archeological record uncovered no evidence of the supposed replacement of the Mousterian (that is, Neanderthal) culture by that of incoming modern populations, the Aurignacians. The more advanced culture evolved *in situ*, he argued.

A powerful but somewhat abrasive personality, Hrdlička failed to persuade his colleagues of what he termed the "Neanderthal Phase of Man." "The only justification for re-opening the problem of the status of Neanderthal man would be afforded by new evidence or new views, either of a destructive or constructive nature," said Grafton Elliot Smith after Hrdlička's major presentation of his unilinear theory in the Huxley Memorial Lecture of 1927. Elliot Smith stood by his own and Keith's assessment of the fossils, declaring, "I do not think Dr. Hrdlička has given any valid reason for rejecting the view that *Homo neanderthalensis* is a species distinct from *H. sapiens*."

New fossils discovered during the 1920s and 1930s in Europe and Asia "were simply assimilated into existing phylogenetic schemes that almost without exception portrayed Neanderthals as an archaic and extinct lineage," notes Spencer. Nevertheless, he explains, these and other later finds were used as evidence by German anatomist Franz Weidenreich to produce "a sophisticated elaboration of Hrdlička's Neanderthal hypothesis involving parallel evolutionary lineages in various regions of the Old World leading through separate Neanderthaloid stages to the modern geographical variances of *H. sapiens*." Weidenreich began his career under Schwalbe at the University of Strasbourg in 1899, which may have made him sympathetic to the unilinear pattern of evolution. Later he worked at the Peking Union Medical College and, as director of the continuing excavations at the Zhoukoudian (Peking Man) cave, enjoyed unique access to the Peking Man material, known then as *Sinanthropus* (equivalent to *Pithecanthropus* and now called *Homo erectus*).

Arguing that Neanderthal anatomy had been misconstrued, Weidenreich complained that in earlier years "it almost became a sport of a certain group of authors to search the skeletal parts of Neanderthal Man for peculiarities which could be proclaimed as 'specialization,' thereby proving the deviating course this form had taken in evolution." According to Weidenreich, an unbiased analysis of the anatomy revealed that the pithecanthropines gave rise to the Neanderthals, which were directly ancestral to modern humans—a scheme reminiscent in broad outline of Gustav Schwalbe's 40 years earlier. Weidenreich's proposal came to be known as the candelabra model of modern human origins: drawn schematically, the long regional ancestries look like an array of candles. Weidenreich's model, elaborated during the 1940s, is the precursor to one of the major positions in the current debate.

Weidenreich was aware that as a consequence of his suggestion that each modern geographical population traces its origins back through Neanderthal and pithecanthropine precursors, modern races might

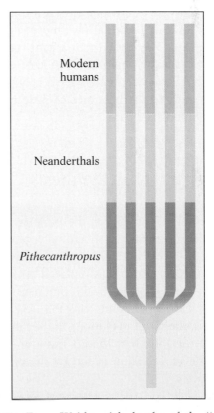

In the 1940s, Franz Weidenreich developed the "candelabra" model of modern human origins, in which geographical populations, established at the Pithecanthropus (Homo erectus) *stage, each passed through a Neanderthaloid stage, to modern* Homo sapiens.

be characterized as having separate origins, even as being separate species. "This is not what I have in mind," he wrote in 1949. "All facts so far available indicate that man branched off as a unit which split afterwards, within its already limited faculties into separate lines. . . . I regard all hominids, living mankind included, as members of one species."

In 1962, however, University of Pennsylvania anthropologist Carleton Coon came close to proposing what Weidenreich had warned against. Coon argued not only that racial differences were ancient, but also that some races had achieved sapienshood earlier than others. "*Homo erectus* evolved into *Homo sapiens* not once but five times, as each subspecies or race living in its own territory passed a critical threshold from a more brutal to a more sapient state," he wrote. The notion that extant racial groups have been genetically separate for at least a million years and that some were relatively recently evolved lent itself readily to the inference of deep biological differences among the races.

Impact of the New Synthesis

Despite Weidenreich's efforts, the unilinear point of view was slow to reemerge. Eventually, however, a confluence of events overturned the hegemony of the presapiens theory, which was replaced with several competing theories, of which the unilinear model was one. The first of these events was the development of the synthetic theory of evolution; the second was the exposure of the Piltdown "fossils" as a hoax; the third was the reevaluation of the La Chapelle-aux-Saints skeleton.

During the late 1930s and early 1940s, evolutionary scholarship was undergoing a revolution that combined the separate sciences of genetics, population biology, and traditional morphology in an evolutionary approach that is sometimes characterized as neo-Darwinism, or the synthetic theory. Julian Hux-

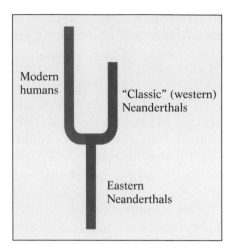

The pre-Neanderthal hypothesis, developed independently in the 1950s by Italian anthropologist Sergio Sergi (the son of Giuseppe Sergi) and American anthropologist Clark Howell, argued that not all Neanderthals should be lumped together. A Neanderthal-like form could have been ancestral both to modern humans and to classic Neanderthals of western Europe, according to the hypothesis.

ley, grandson of Thomas Henry Huxley, published *Evolution: The Modern Synthesis* in 1942—a landmark text. Others instrumental in forging the revolution were Theodosius Dobzhansky, Ernst Mayr, and George Gaylord Simpson. While paleoanthropology had traditionally been a descriptive, typological activity, the synthetic theory encouraged a more evolutionary, populational approach. This approach minimizes peculiar anatomical features any individual specimen might possess and emphasizes the importance of anatomical variation within populations. A meeting at Cold Spring Harbor, New York, in 1950 titled "Origin and Evolution of Man" formally marked the marriage between the synthetic theory and paleoanthropology.

One consequence of the new thinking was the development of what came to be known as the pre-Neanderthal hypothesis, an idea with roots in the 1920s that was only fully articulated in the 1950s. Italian anthropologist Sergio Sergi (the son of Giuseppe Sergi) and American anthropologist Clark Howell separately argued that in an evolutionary context it was inappropriate to lump all Neanderthals together. The increasing body of discoveries revealed that the anatomy of Neanderthal populations in western Europe was unquestionably more exaggerated than that of eastern populations, yet the former are usually referred to as "classic" Neanderthals. Perhaps, suggested Sergi and Howell, some of the less extreme specimens—from Europe and the Near East—represented populations that were ancestral to both the "classic" Neanderthals and modern humans. These supposed pre-Neanderthalers included finds from famous sites like Swanscombe, Steinheim, Fontechevade, Kaprina, Teshik Tash, Tabun, and Skhul. In effect the pre-Neanderthal hypothesis rescued some Neanderthal-like populations from evolutionary oblivion—while leaving the classic Neanderthals to that fate.

Forgery Exposed, Forger Unknown

When the Piltdown fossils were first presented to anthropologists in 1912, they appeared to offer evidence of a modern form early in human prehistory, thus conforming with prevailing theoretical ideas. Although the presapiens theory that arose from the analysis of these fossils persisted, tangible support for it remained elusive. A second Piltdown specimen, discovered in 1915, served in some cases to remove skepticism about the identity of the original fossil. But no more examples came to light: no more instances of modern human form early in human prehistory.

Gradually the Piltdown fossils came to be regarded as something of an enigma, isolated, increasingly at odds with other evidence. By the early 1950s, many pithecanthropines (*Homo erectus*) had been found; they were distinctly more primitive than Pilt-

down and yet more recent in age. Moreover, australopithecine fossils had been unearthed in South Africa. Acknowledged as early members of the human family, the australopithecines were apelike in many respects and yet were thought to be little older than Piltdown Man, a very modern form. The paradox was resolved in 1953 when Joseph Weiner, an anatomist at Oxford University, showed the Piltdown fossils to be forgeries. The skull fragments were relatively modern, while the jaw material was from an ape—an orangutan, as it was later demonstrated.

The author of the forgery remains a mystery, although every major figure involved in the discovery and study of the fossils has, over the years, been proposed as the perpetrator, including Charles Dawson, who died in 1916. No direct evidence to solve the riddle unequivocally has ever come to light; accusers have been forced to rely on circumstantial evidence, never universally convincing. The demonstration that Piltdown was a fraud, however, removed it as a factor in the body of human fossil evidence.

An Insight into Boule's Motives

Although by the early 1950s Piltdown was lost as support to proponents of the presapiens theory, the second major pillar of the theory—the bizarre anatomy of the Neanderthals—remained intact. Anthropologists continued to espouse the view that "the European Neanderthals were a specialized and extinct branch of the human evolutionary tree," says Frank Spencer. "During the 1950s the presapiens view was promoted largely through the efforts of Henri Vallois, Boule's successor at the Musée de l'Homme and Institut de Paleontologie Humaine in Paris." Louis Leakey, whose mentor had been Arthur Keith, also supported the theory at this time and essentially continued to do so until his death in 1972. From the mid-1950s onward, however, this second pillar of the presapiens theory also began to crumble.

Anatomy of a Forgery

Piltdown Man was one of the most audacious of scientific frauds. Technically poor in execution, the "fossil" material nevertheless was accepted by many prominent anthropologists as important evidence of a modern form of man in the Lower Pleistocene, although as more early human fossils were discovered in Asia and Africa, Piltdown Man, with his modern cranium allied with an apelike jaw, became increasingly anomalous in the unfolding hypotheses of human prehistory.

At a gravel deposit near the village of Piltdown, Sussex, nine fragments of cranium were ultimately unearthed, forming much of the left side of the humanlike braincase. Part of the right side of an apelike jaw was also discovered, which bore two molar teeth but lacked the chin part and the joint that would have articulated with the skull. The socket for the canine tooth was present, but the tooth itself was missing. (Later a large canine was discovered in the deposits, which fit neatly in the socket, accentuating the apelike appearance of the jaw.)

Also discovered at the site were mammalian fossils—including beaver, red deer, and horse, which indicated an Early Pleistocene age—and some crude stone tools. Charles Dawson said that the gravel layer was part of a sequence well known in the region, known to be Early Pleistocene and thus consistent with the age of fossil fauna. If the human remains were also Early Pleistocene, they would be older than any hominid fossil known at the time—particularly *Pithecanthropus* and Neanderthal Man; moreover, the modern form of the cranium and the implied large brain size revealed a more advanced evolutionary stage than these fossils. Announced to the anthropological world in December 1912, the Piltdown fossils were named *Eoanthropus dawsoni* (Dawson's Dawn Man).

Several points of dispute surrounded the Piltdown fossils. The most important questioned whether the cranium and the jaw had belonged to the same individual or were the separate remnants of a human and an ape that had fortuitously become associated in the same deposits. The majority of British anthropologists accepted that the fossils represented a single individual, while the prevailing sentiment in continental Europe and the United States was that Piltdown Man was a chimera of human and ape. When, in 1915, Dawson reported that he had discovered two cranial fragments and a molar tooth from a second site two miles from Piltdown, the British position seemed strengthened: the form of the cranial bone and the shape of the tooth were just like those of the first Piltdown individual.

The unmasking of the hoax was a slow process, begun with questions about the age of the Piltdown fossils. In 1935 it was realized that Dawson had erroneously claimed an Early Pleistocene age for the Piltdown gravel layer; it was instead Late Pleistocene. This would make Piltdown Man much younger than had been supposed and, the apelike jaw aside, would fit more comfortably with prevailing theories. But there were other implications to the revelation. If the gravel layer was indeed Late Pleistocene, how was the presence of Early Pleistocene fossil fauna to be explained? Clearly, they must have come from somewhere else—in which case, the issue of the association of humanlike cranium and apelike jaw could once again be raised.

Kenneth Oakley (1911–1981), an anthropologist at the British Museum (Natural History), began a research program designed to solve the mystery of the age of the fossils. He developed a method of dating fossils using fluorine content. Fossils absorb fluorine from the soil in which they are buried, and the longer they are in the ground the more fluorine they accumulate. Oakley's method required demanding chemical analysis and was further complicated by the fact that different soils contain different amounts of fluorine, affecting the absolute quantity of the chemical that fossils might take up. Comparisons of fossils from different sites, therefore, were unreliable, but comparisons within a single site were valid. By the late 1940s Oakley had satisfied himself that the method worked and could be applied to the Piltdown fossils, which he did in 1949.

The results were unequivocal. Some of the fossil fauna were Early Pleistocene (with 2.3 percent of fluorine), as had been surmised from their evolutionary stage, while the Piltdown fossils were clearly Late Pleistocene to Recent, with a mere 0.2 percent of fluorine. Unfortunately the method was insufficiently sensitive to discern any age differences among the fossil fragments of the Piltdown Man. Oakley did note that when he was drilling the teeth to obtain material for the chemical test, the dentine beneath the dark brown surface of the fossil was pure white, just like "new teeth from the soil." Bone and teeth become stained as they rest in sediments, but in genuine fossils the color extends through the material.

There the matter rested until July 1953, when the Wenner–Gren Foundation organized a meeting of anthropologists in London. A conversation between Oakley and Joseph Weiner (1915–1982) prompted the latter to think the unthinkable: that the Piltdown fossils had defied explanation because they were fraudulent. The following day, in company with fellow anthropologist Wilfrid Le Gros Clark (1895–1971), Weiner examined casts of the Piltdown fossils and immediately discerned evidence of fraud. The evidence was in the wear pattern of the teeth. Had it been natural, the first molar would have been

more worn than the second, but it was the same. The wear on the two molars should naturally have formed a flat plane, but they were offset from each other; microscopic examination revealed crisscross scratches, as if the teeth had been worn down by an abrasive.

If it was so obvious, why had no one seen signs of deception earlier? "They had never been looked for," wrote Le Gros Clark. "Nobody previously had examined the Piltdown jaw with the idea of a possible forgery in mind, a deliberate fabrication."

Using a more sensitive fluorine method, Oakley was later able to show that the cranial fragments were Late Pleistocene, while the jaw material was relatively modern. The color of the "fossils" was shown to be the result of staining with a paintlike substance, possibly Vandyke brown. In 1982 Jerrold Lowenstein, a biomedical researcher at the University of California, San Francisco, used sensitive immunological methods to reveal that the jaw was that of an orangutan. The forger(s) have never been identified with certainty.

The famous John Cooke painting of the Piltdown Men. Center, *wearing a laboratory coat, is Arthur Keith;* to his left *(seated) is William Plane Pycraft and Edwin Ray Lankester;* to his right, *Arthur Swayne Underwood.* Behind *(left to right), Frank Orwell Barlow, Grafton Elliot Smith, Charles Dawson, and Arthur Smith Woodward.*

With Boule's vision of Neanderthals as bent-knee, bent-hip, stooping gaited cavemen still current in the mid-1950s, anatomists William Straus of the Johns Hopkins University and A. J. E. Cave of St. Bartholomew's Hospital Medical College, London, reexamined the La Chapelle-aux-Saints skeleton. They observed that this Neanderthal probably had walked with something of a stoop, but that this was due to severe arthritis in the vertebral column. The neck vertebrae were not like those of a chimpanzee, as Boule had contended, nor was the pelvis apelike in any way, and the feet bore no signs of being prehensile.

So impressed by the modern anatomical appearance of the skeleton were Cave and Straus that in their 1957 report of the reanalysis they said: "If he could be reincarnated and placed in a New York subway—provided that he was bathed, shaved and dressed in modern clothing—it is doubtful whether he would attract any more attention than some of its other denizens." The two anatomists were indulging in a degree of hyperbole; modern though Neanderthals may be in many respects, their skeletal robusticity and strikingly protruding midfacial region would surely attract attention as being unusual, even on the IRT.

The rehabilitation of Neanderthal was effectively completed by Loring Brace, whose 1964 paper "The Fate of the 'Classic' Neanderthals" was widely influential. In it he noted that "interpretation of the hominid fossil record has inevitably been colored by the climate of opinion prevalent at the time of the discovery of the major pieces of evidence"—an observation whose verity is repeatedly affirmed in this volume. Brace, reexamining the La Chapelle-aux-Saints skeleton, concluded that Boule had described anatomical features that simply were not present. "There is no trace of evidence that Neanderthalers had exceptionally divergent great toes or that they were forced to walk orang-like on the outer edge of their feet; there is no evidence that they were unable fully to extend their knee joints; there is no evidence that

their spinal columns lacked the convexities necessary for fully erect posture; there is no evidence that the head was slung forward on a peculiarly short and thick neck; and there is no evidence that the brain was qualitatively inferior to that of modern humans." (This last observation was later confirmed extensively by Ralph Holloway, an anthropologist at Columbia University, who is among the world's experts on modern and fossil hominid brains. At a 1984 gathering of anthropologists in New York, he stated that his study of the putative primitive characteristics of Neanderthal brains showed them to be "simply nonexistent.")

Brace was effectively corroborating and extending Cave and Straus's earlier work, again pointing out that Boule, an eminent anatomist, had apparently ignored the effect of pathology on the skeleton. Take account of the pathology and it was clear that the Neanderthal spine indeed possessed the kind of alternating curves that produce upright posture in modern humans. But Boule had disregarded this factor, choosing to describe the anatomy in a way that distanced Neanderthals from modern humans. "I have been able to find absolutely no indication, or even a hint, that Boule fudged or fraudulently mis-represented his research," concludes Michael Hammond. "Boule sincerely looked upon science as 'one of the principal sources of happiness' when 'guided by that interior flame,' the 'love of Truth.' His La Chapelle-aux-Saints reconstruction was his most important contribution to this search; and he believed that the best scientific techniques available had been used, and that his conclusions were dictated by the data themselves."

If Boule was neither mendacious nor stupid, how can his mischaracterization be explained? Like all scientists, as Brace observed, Boule was guided by a prevailing theory about how the world works. In this case, his conceptual viewpoint concerned the pattern of a species' evolutionary history, and it was shaped through his association with Albert Gaudry, his teacher, friend, and patron at the Museum of Natural History. Boule succeeded Gaudry as professor of paleontology at the museum in 1902. In their collaborative work on a wide range of fossil mammals, the two men had repeatedly found branching patterns, implying that many species stopped short of contributing to the final history of their groups. "La Chapelle-aux-Saints capped these studies by demonstrating that in the family of man, species like the Neanderthals, which were previously thought to be ancestral were in fact collateral. They represented dead ends," observes Hammond. In other words, because of his experience with other mammalian groups, Boule expected to find evidence for a treelike evolutionary history for humans; so he did.

The conceptual approach to evolutionary patterns developed by Gaudry and followed by Boule was a challenge to the prevailing doctrine of a unilinear evolutionary history, the chief supporter of which was Gabriel de Mortillet (1821–1898). An archeologist and paleontologist of renown, Mortillet developed the idea that a single great glacial period had been responsible for the evolutionary transformation of Neanderthals into modern humans. In parallel he outlined a scheme for cultural evolution over the same period. Both schemes proved to be far too simple, and Boule became a sharp critic of their geology, archeology, and anthropology.

In 1888, Boule attacked the glacial geology in Mortillet's model, explains Hammond, "and over the next years he described Mortillet's theoretical scheme as 'a mirage of doctrines' and a 'mummy which he encircles every day with new bandages' in order to protect it from scientific criticism." Mortillet died a year later. Boule, in company with his friend the Abbé Breuil, then systematically took apart their deceased foe's ideas on archeology. "This meant that by 1908, all that was left of Mortillet's classification was the linear paleontological ladder, and this Boule demolished." Evolutionary change, Boule argued, "is not accomplished as simply as was believed in the beginnings of the science; that unilinear series appear to us as more and more rare; and if they exist, it is extremely difficult to find them or to pursue them for any length of time."

The expulsion of Neanderthal from the direct lineage of human history by Boule was therefore part of the dismantling of a widely influential evolutionary research program in late-nineteenth-century France. It was also an example, as Brace noted, of how theoretical preconceptions color the interpretation of evidence, a phenomenon common to all science but particularly strong in paleoanthropology.

Single Species Hypothesis Expires . . .

By the late 1960s Neanderthals had been restored to, in many people's eyes, their rightful place: as direct ancestors of modern humans. The unilinear theory was at last successfully revived, now as one of a handful of competing theories: these included the pre-Neanderthal hypothesis and a version of the presapiens hypothesis. Unquestionably Loring Brace's paper "The Fate of the 'Classic' Neanderthals" had played the key role. "I suggest that it was the fate of the Neanderthal to give rise to modern man, and, as has frequently happened to members of the older generation in this changing world, to have been perceived in caricature, rejected, and disavowed by their own offspring, *Homo sapiens*," he concluded. Fossils that had been discovered in Europe and Asia during the first half of the century were now interpreted by Brace and his supporters within the unilinear theory as evidence of evolution toward *Homo sapiens* in many different parts of the Old World.

Brace argued that the adaptive niche of humans was "a cultural niche." Other scholars had "failed to appreciate the fact that culture, rather than climate, has been the prime factor to be reckoned with in assessing selective pressures operating on man." According to Frank Spencer, "Brace's hypothesis is an

Left: *The discovery in 1975 of the* Homo erectus *skull numbered KNMER 3733 forced Loring Brace and Milford Wolpoff to abandon their Single Species hypothesis. The skull was contemporaneous with the extremely robust* Australopithecus boisei *skull KNMER 406 (*right*). The anatomical differences between the two were too great to be accommodated by intraspecies variation, which the Single Species hypothesis would have demanded.*

elaborate restatement of Hrdlička's suggestion that changes in masticatory function were an integral process in later patterns of hominid cranial function. . . . [W]ith the introduction of more specialized tools in the Middle and Upper Paleolithic to perform tasks previously handled by the anterior teeth there was a progressive relaxation in the amount of stress generated in the craniofacial skeleton, which ultimately precipitated a series of morphological changes that signalled the appearance of the modern cranial configuration.'' In other words, similar kinds of anatomical modernization were going on in populations throughout the Old World as a result of the common

adoption of a more advanced form of technology. And in a common technological context, anatomical variation loses some of its former significance.

According to Brace's so-called Single Species hypothesis, only one species of hominid existed at any given period in human evolution—the ultimate expression of the unilinear pattern. Brace was adding an extra stage to Schwalbe's original, three-stage scheme; it now read: australopithecines, pithecanthropines, Neanderthals, modern humans. Milford Wolpoff, also at Michigan, joined Brace as a vigorous spokesman for the hypothesis. Both men argued that the degree of anatomical difference displayed among populations of fossils from each slice of prehistoric time could be encompassed within intraspecies variation, from the beginning of human evolution to the present day.

Although Brace had persuaded many that a unilinear pattern of evolution had prevailed in recent human prehistory, few accepted the notion for the entire period of human evolution. Most anthropologists interpreted the breadth of anatomical variation evident in the fossils as indicating the coexistence of several species. As we have seen, at least three—and perhaps twice that number—hominid species existed some 2 million years ago, for instance. By the mid-1970s Brace and Wolpoff were virtually the sole adherents of the Single Species hypothesis for all of human evolution. Only with the discovery of a *Homo erectus* cranium from the Lake Turkana region of Kenya in 1975 did the Michigan scholars abandon their hypothesis. The cranium was large (brain capacity approaching 900 ccs), the face was short, and the cheek teeth small. Pertinent was its discovery in deposits of the same age—about 1.5 million years—from which a robust australopithecine had been

taken. The latter's small brain (about 450 ccs), protruding face, and enormous cheek teeth showed it to be dramatically different from the *Homo erectus* cranium. No single species could accommodate such variation, Brace and Wolpoff finally admitted.

. . . But Not Completely

Wolpoff, now a major protagonist in the current debate on the origin of modern humans, nevertheless insists that the unilinear hypothesis holds for the later stages of human prehistory. A scientific tradition carrying the names of Schaaffhausen, Schwalbe, Hrdlička, Weidenreich, and Brace therefore persists. In the following chapter we will see how the modern version of this tradition measures up against the modern version of denying direct ancestry between Neanderthals and modern humans. And we will see how close to completing the third flip modern opinion has come.

3

TWO MODELS

The 1990 annual gathering of the American Association for the Advancement of Science offered the usual wide range of intellectual fare, from the latest discoveries in neurobiology and concerns over the loss of biodiversity to the mysteries of the early universe and of modern economics. Before a packed audience on the midmeeting Sunday afternoon, scientific scrutiny was focused on the fossil evidence for modern human origins as presented by paleoanthropologists who were most familiar with the limitations inherent in this source.

The program appeared to promise a balanced assessment of the fossil data, characterized as "the only direct evidence with the power of refutation." In fact, the half-dozen speakers organized by the University of Michigan's Milford Wolpoff consistently reiterated a single point of view, declaring the opposing interpretation demolished by their evidence. There is no absolute requirement, of course,

A Cro-Magnon skull (Cro-Magnon I), found in 1868 at Les Eyzies, France, and dated at about 30,000 years old.

that scientific symposia balance equal opportunities for proponents on both sides of any question. In the context of general scientific gatherings like the AAAS, however, one-sidedness is unusual. "It was meant to be a sales pitch," Wolpoff later admitted to a reporter for *Discover* magazine; "we planned the whole thing, rehearsed it, worked over the exact phrasing." He explained that the scientific ideas behind his group's Multiregional Evolution hypothesis

were complicated and, as a result, had not been well communicated to a wider public. This symposium was structured to remedy that problem—not so much to persuade other anthropologists of their view as to air it publicly.

If the content of the scores of the following day's newspapers, including *The Times* of London, is any measure, Wolpoff and his colleagues achieved their aim. "A group of specialists on fossils said that

Milford Wolpoff, of the University of Michigan, is a strong proponent of the Multiregional Evolution hypothesis.

studies of skulls and other remains of humanlike creatures found in Asia and Europe showed that the Garden of Eden theory must be wrong," said one account. (The Garden of Eden theory—also known as the Out of Africa hypothesis—is one phrase used to describe the theory Wolpoff and his colleagues were attacking so vigorously.) "Fossils are the real evidence," one of the speakers was quoted as saying in another article, "and they don't support this theory." A third story declared the Garden of Eden theory "under attack this week at the nation's top annual gathering of scientists."

Scientific debates are never settled by the weight of public opinion, of course. Only the accumulation of a body of irrefutable evidence can do that. Nevertheless, some issues are inherently of greater interest than others to a wide audience: the origin of modern humans naturally captures the public's attention. The language of the debate, moreover, may reflect underlying passions. References to the Garden of Eden and invocations of "killer Africans sweeping across Europe and Asia" make it clear to observers that more than dry scientific facts are at stake. A theory of modern human origins is, in effect, a description of how our immediate ancestors behaved. Questions of race and racial differences are implicit here, so it is perhaps inevitable that the discussion should be both voluble and passionate at times.

In recent years, as we have seen in previous chapters, new scientific evidence has been forthcoming, some concerning the fossils themselves and some from genetics, a relatively novel source of insight for anthropologists interested in the later stages of human prehistory. Science thrives on new evidence, which, however, comes with no guarantee of consensus. The effect in this case has been both to clarify and to polarize opposing positions, effectively eliminating the middle views. This is where the debate over modern human origins now finds itself. Two mutually exclusive arguments currently demand adherence.

Three sources of evidence are adduced to support and test these arguments. The fossils themselves, carrying anatomical characteristics that, in principle, should be interpretable in terms of evolutionary relationships through time and space, will be the topic of the present chapter. The molecular vestiges of the evolutionary history that each of us carries genetically are the subject of the following chapter. Chapter 5 will examine the behavioral evidence: technology manufactured, clues to subsistence, different modes of life. We will begin to see how surprisingly difficult it is to define the central core of our interest: what is meant by "modern humans."

Four Hypotheses

"Twenty years ago, when I embarked on my doctoral research, there were four main hypotheses concerning the origin of modern humans," recalls Christopher Stringer, based at the Natural History Museum, London. One was the Neanderthal phase (unilinear) hypothesis, elaborated most recently by Loring Brace, as we saw in the previous chapter. The second, known as the spectrum hypothesis, varied this by postulating much greater interaction between different populations, which blurred the distinctions between them. Third, the pre-Neanderthal model of Clark Howell and Sergio Sergi saw an early, generalized Neanderthal population as having diverged relatively recently to produce the classic Neanderthals of western Europe—which eventually became extinct—and modern humans. Fourth was the presapiens model, going back to Marcellin Boule but continued by his countryman Henri Vallois and by Louis Leakey. In this model an ancient split is envisaged between the true *sapiens* lineage leading to modern humans and the lineage leading to other groups such as *Homo erectus* and the Neanderthals.

One model Stringer did not address in his doctoral research was the idea of a single center of ori-

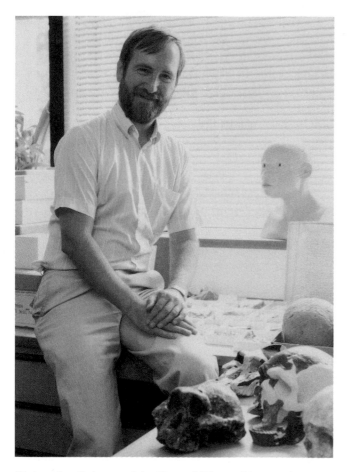

Christopher Stringer, of the Natural History Museum in London, is a strong proponent of the Out of Africa hypothesis.

gin for modern humans, followed by migration into the rest of the Old World. William Howells, of Harvard University, was the scholar most closely associated with this hypothesis, which was incompletely formulated at the time, not least because of the rather sparse fossil record and uncertainties of dating. Yet two decades later, this has become one of the two dominant hypotheses, with Africa as the center of origin and Stringer as its chief proponent, at least from the point of view of fossil evidence.

Stringer describes the four principal views of the day as "straitjackets," as constraining on research as the poor fossil record and the uncertain chronological evidence. Another straitjacket was "the conviction on the part of many workers that technological change was the only possible mechanism behind the evolution of modern humans." The reason is easy to appreciate. Europe has long dominated the archeological record of these later stages of human prehistory, partly because of selective preservation, but mostly because of its long tradition of exploration and research. Moreover, as we shall see in greater detail in Chapter 5, the European fossil record appeared to contain a clear signal of the emergence of modern humans: the Middle to Upper Paleolithic transition.

Origin of the Out of Africa Hypothesis

About 40,000 years ago, the long-established and relatively limited flake technology of the Middle Paleolithic was suddenly replaced by an exquisitely refined and rapidly evolving blade technology of the Upper Paleolithic, which also included extensive use of bone and antler as raw material. In addition, evidence of body ornamentation began to emerge, as did so-called artistic expression on utilitarian objects and cave walls. The Upper Paleolithic people—modern humans—had apparently burst on the scene with dramatic effect, leaving the less refined Middle Paleolithic people behind. European researchers naturally read all this as evidence of the origin of modern humans—an event few were surprised to learn had occurred on their own continent. Today's scientific opinion is different.

"If it is the *origin* of modern humans you are interested in, then Western Europe is clearly something of a backwater," observes Fred Smith, an anthropologist at Northern Illinois University. "One thing that is becoming clear is that the real action was elsewhere." In the same vein, Arthur Jellinek of the University of Arizona recently reflected on the impetus for finding prehistoric sites: "I wonder what our view of the evolution of modern humans would have been like if the industrial revolution and the intellectual curiosity that followed it had started in China or Java or India, with just a few aberrant fossils from the geographical backwater of Western Europe." The archeological signal in Europe 40,000 years ago is strong and clear, but almost certainly it indicates an event in the history of modern humans other than their origin. History unfolded as it did, however, and the Eurocentric archeological view has only recently been thrown off. The thrall of dramatic technological change, as a necessary mark of the origin of modern humans, remains to some extent influential.

The hypothesis Stringer now holds—that of a recent, single origin of modern humans, in Africa—had a long genesis. "I was inspired by Brace's polemical article in *Current Anthropology*, in 1964, to feel that the Neanderthals had indeed been unfairly treated by much previous research," he recalls of his first foray into the topic. He therefore turned to multivariate analysis, a statistical approach to comparing anatomy and inferring relationships based on the analysis of scores of measurements of a series of fossil traits. The approach was much in vogue at the time, not least because statistical manipulation carried a greater air of powerful, objective scientific method than judgments of anatomical similarities based on direct observation alone. "I felt, along with many others, that multivariate analyses could give us the objectivity we were seeking," says Stringer. "I had only to feed the data into the computer, and the answers would rapidly emerge!"

Stringer set out on a three-month trip around Europe in 1971 to gather the data he would feed into his computer. Skulls collected from Europe, the Near East, and North Africa were his target. "In most places I was received with great courtesy, despite my lack of status and my unconventional appearance," he recalls, "but nevertheless I found some important material completely unavailable for study." This was usually for legitimate reasons, such as a specimen being on loan elsewhere for study. But one curator at a major museum in Paris tried to prevent Stringer from seeing a particularly important specimen, claiming it was temporarily at another laboratory. Knowing this to be false, Stringer was able to examine the material only by recourse to subterfuge and the aid of a sympathetic researcher in the same institution. Of such acts is the unusual reputation of paleoanthropology made.

When Stringer's computer analysis was completed, answers emerged, but no clear conclusion. In spite of the exaggerated development of various aspects of anatomy in the Neanderthals of western Europe, the statistical analysis detected echoes of similar exaggeration in some Neanderthal-like specimens in the Near East: Amud, for instance, in Israel. Other Near Eastern specimens, from the cave of Skhul, for example, were much more like the Upper Paleolithic people of Europe. One clear signal, however, was the morphological gulf between the Neanderthals and the Upper Paleolithic individuals. As a result, Stringer felt he could reject the simple unilinear hypothesis of Brace, which required that Neanderthals evolved into Upper Paleolithic people. When he looked at specimens earlier than Neanderthal, moreover, such as Swanscombe (from England) and Fontechevade (from France), they appeared to be too archaic or Neanderthal-like to support Vallois and Leakey's presapiens hypothesis; that hypothesis demanded the existence of modern forms early in the record, and Stringer could find none. With the unilinear and presapiens models eliminated, the spectrum hypothesis and the pre-Neanderthal hypothesis remained as possibilities.

Importance of a Good Date

"Looking back at my thesis research . . . , it is obvious that it suffered from naïveté about the power of multivariate methods to resolve evolutionary problems," observes Stringer three decades later. Although still utilized by physical anthropologists, and considered useful in direct, limited anatomical comparisons, multivariate analysis has proven less powerful than expected in the search for subtle evolutionary patterns. But there were other problems too, not least of which was the poor, and poorly dated, fossil record with which he had to deal. Both better dates and new specimens emerged over the years, making a difficult job somewhat easier. As significant, however, was the emergence of new techniques for comparing anatomy.

For some years a small group of biologists had been developing a method known as cladistics, a potentially powerful tool for discerning relationships among species by sorting out those anatomical similarities that are significant from ones that are not. The technique depends on the identification of anatomical characters that exclusively unite a group of species as evolutionarily related. During evolution novelties arise from time to time. The species in which the new character arises, and all the descendant species, uniquely share the novelty and are known as a clade; the novelty is termed as a shared derived character for the group.

An example is the possession of fingernails, which unites primates as descendants from a single common ancestor. The character is therefore useful for identifying primates, but since all primates have fingernails, the character cannot discriminate among different groups of primates, such as anthropoids from prosimians. A novelty that arose *within* the primates—the bony ridges above the eyes, for instance—identifies all hominoids (apes and humans) as derived from a single ancestor within the primate clade. The possession of a bipedal mode of locomotion further identifies hominids as derived from a sin-

gle ancestor within the hominoid clade. In the 1970s anthropologists were starting to apply the cladistic technique to issues of evolutionary relationships among hominid fossil specimens.

"Through my contact with workers such as Peter Andrews, Eric Delson, Jeff Schwartz, and Ian Tattersall, I was becoming aware of the potential of cladistics to contribute to my understanding of the data I was trying to analyze," recalls Stringer. As a result, the specimens began to arrange themselves into three clearly defined groups. One was the Neanderthals, both early and late, from Asia and Europe. A second was modern, including the Upper Paleolithic people, two samples from Israel (Skhul and Qafzeh), and one from Africa (the Omo 1 cranium from Ethiopia). A third group was more primitive, including the Petralona skull from Greece, Broken

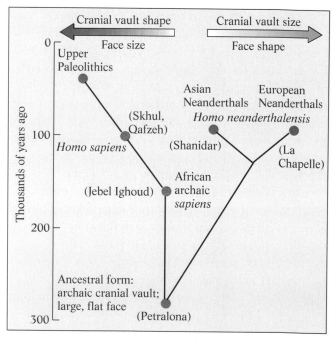

The evolution of cranial shape produced three patterns in the later stages of human prehistory. The Neanderthal pattern diverged from that of modern humans, making an ancestral relationship between them unlikely.

Border Cave, in southern Africa, is the site of some of the oldest anatomically modern human fossils (skull fragments), which may date to 80,000 years ago or more.

Hill from Zambia, and Arago from France. Most important, however, a pattern began to emerge concerning changes in the configuration of cranium and face shape.

Before half a million years ago, the overall shape of the head conformed to a single pattern: the cranial vault was somewhat long, broad, low, and of small volume; the face was broad, large, and flat. This can be thought of as the ancestral configuration. Between half a million and 50,000 years ago, three different patterns emerged. In the first, which led to modern humans, the cranial vault became higher, narrower, and shorter, while the face retained the same overall shape on a smaller scale. In the Neanderthals, the reverse happened; cranial shape kept the primitive pattern, though expanded in volume, while the face became modified—narrower at the top and distinctly projecting in the middle. A third group, retaining much of the overall primitive shape, included a famous fossil from North Africa, Jebel Ighoud. Stringer concluded that the Skhul/Qafzeh sample and the Asian Neanderthals seemed to be veering in opposite directions: toward modern humans and European Neanderthals, respectively, with different evolutionary fates.

There seems to have been no genetic continuity between the archaic populations (including the Neanderthals) in Europe and the Near East and the modern humans that eventually became established there. In other words, the early populations in that region of the world were not the ancestors of the people who lived there later. What of other regions?

The fossil evidence outside of Europe is sparse, but from what there is Stringer could see no signs of continuity from older populations to modern populations in any of them—apart from Africa. In Africa, it was becoming apparent that certain specimens with modern human characteristics (such as the Omo 1 specimen, fragments from the Klasies River Mouth Cave in South Africa, and Border Cave) were among the oldest anywhere in the world: very probably, the oldest.

1 Products of radioactive decay interact with nearby atoms, boosting energy levels.

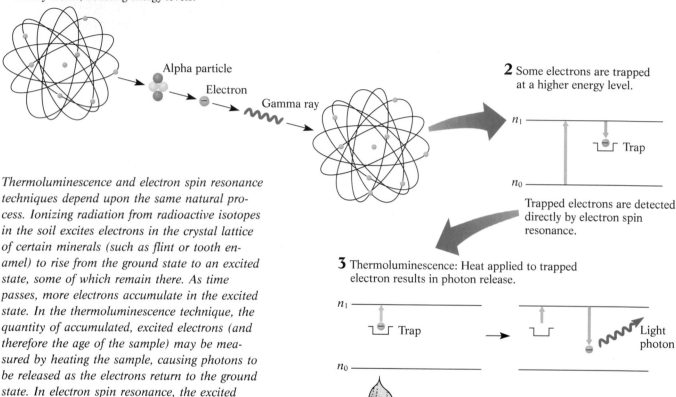

Alpha particle

Electron

Gamma ray

2 Some electrons are trapped at a higher energy level.

n_1

Trap

n_0

Trapped electrons are detected directly by electron spin resonance.

3 Thermoluminescence: Heat applied to trapped electron results in photon release.

n_1

Trap

Trap

Light photon

n_0

Thermoluminescence and electron spin resonance techniques depend upon the same natural process. Ionizing radiation from radioactive isotopes in the soil excites electrons in the crystal lattice of certain minerals (such as flint or tooth enamel) to rise from the ground state to an excited state, some of which remain there. As time passes, more electrons accumulate in the excited state. In the thermoluminescence technique, the quantity of accumulated, excited electrons (and therefore the age of the sample) may be measured by heating the sample, causing photons to be released as the electrons return to the ground state. In electron spin resonance, the excited electrons are measured more directly, by applying a magnetic field and microwave irradiation.

Uncertainties in dating these specimens have plagued their interpretation for years. Recently, however, dating techniques such as thermoluminescence and electron spin resonance have been applied with some success. Both techniques depend on natural radioactivity in the soil, which boosts electrons to higher energy levels in "target" materials such as flint or tooth enamel. The longer such material is buried, the greater will be the accumulation of boosted electrons, thus providing a measure of the passage of time since burial. The thermoluminescence technique measures the higher energy electrons by the release of light that occurs when the target material is heated. Electron spin resonance detects the high-energy elec-

trons more directly. As a result of new dates from these techniques, confidence is building that some of the early anatomically human specimens in southern Africa are close to 100,000 years old, making them contemporaries of Neanderthals in Europe and Asia.

A decade ago, however, Stringer had to base his assessment on dates from Africa that had wide margins of uncertainty. In 1982, at a major anthropological conference in Nice, he publicly presented an African origin model for the first time. The hypothesis rested on three points. First, the evidence of an early appearance of anatomically modern humans in Africa, notwithstanding its uncertainties. Second, the virtual absence of any similar evidence from Europe and

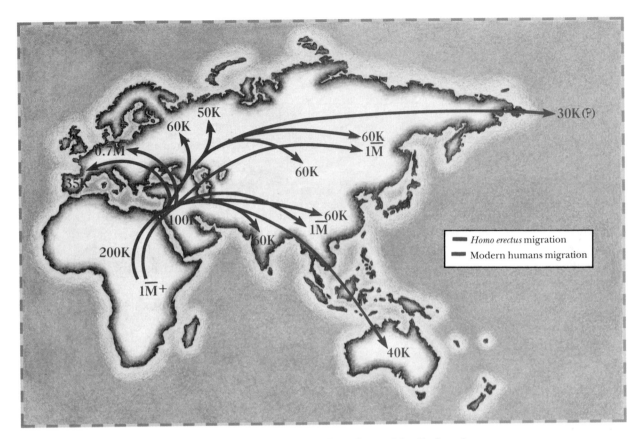

A map of migration out of Africa can be assembled, based on dates of fossils found in various parts of the Old World. Two migrations can be discerned: that of Homo erectus, *about 1 million years ago; a second, beginning 100,000 years ago, of modern human populations.*

Asia. And third, an assumption that the Skhul/Qafzeh fossils from the Near East were of relatively recent origin, no more than 50,000 years old. The hypothesis was published at the end of 1982 in *Current Anthropology.*

By this time others had begun to think along the same general lines, though none as explicit or as exclusive as those framed by Stringer. Researchers such as Desmond Clark, Reiner Protch, Peter Beaumont, Philip Rightmire, and Gunter Bräuer had all pointed to Africa as the site of origin of modern humans, but each included considerable input from other regions in various ways. Stringer did not sug-

gest that the modern humans from Africa spread out into the rest of the world, completely replacing all existing populations of premodern people; such a suggestion would later come from geneticists. But he saw interbreeding between incoming and established populations as relatively limited. In any case, the so-called Out of Africa model was established: the notion that anatomically modern humans arose somewhere in sub-Saharan Africa, followed by population spread into the rest of the Old World, where there might have been some degree of gene flow through interbreeding with preexisting populations of archaic *sapiens* people.

The Out of Africa model is simply one of pattern, insists Stringer; it does not prescribe process or biology. "The probability that a speciation event occurred seems to me to be high," he says, "but it is not a sine qua non of the model." There seems to have been an early appearance of anatomically modern humans in Africa by some 100,000 years ago, followed over the next 70,000 years by a disappearance of archaic forms and the establishment of these modern forms in the rest of the world. Whether this change was the result of newly evolved language, superior technology, new forms of social organization, or sharper cognitive skills that gave an edge to *Homo sapiens* over archaic *sapiens* people, the Out of Africa model does not specify. Nor need it specify such elements in order to be valid. "Those who demand to know what allowed the replacement to occur, and what factors drove the dispersal event, before they will accept that it occurred at all, are asking the wrong sorts of questions," argues Stringer. "For over a century, scientists have accepted that the dinosaurs rapidly became extinct, with the real causes only now becoming apparent. Equally, we accept that bipedalism evolved in the hominid lineage over 4 million years ago, without any clear idea of how and why this happened."

Feasibility of Gene Flow

Some critics of the model do, however, air assumptions about what must—or simply could not—have happened. Geoffrey Pope, for instance, of the University of Illinois, said at the 1990 AAAS meeting that if modern humans from Africa had "invaded" Asia, then a "Rambo-like technology" would have suddenly appeared in the record. "There is no indication of the rapid appearance of new tools of any kind . . . in Pleistocene Europe or Africa," he said. "If invading anatomically modern *Homo sapiens* brought with them new technological innovations, then they left no trace in the archeological record of the region they colonized." Similarly, Geoffrey Clark of Arizona State University balks at the idea of large population movement. "I simply do not believe that the physical migration of peoples played a significant role in human macroevolution," he said recently. In fact, the term migration as it is often used in this context is probably misleading, insofar as it conjures up the idea of people setting out on a journey with the intention of colonizing new and distant lands. More likely, population expansion and spread occurred slowly and fitfully, following ecological opportunities—just as has happened many times in other species. It is worth noting the simple calculation that a modest 20-mile-per-generation "migration" could carry a population from sub-Saharan Africa to western Europe in a mere 10,000 years.

Milford Wolpoff also makes assumptions about what "must have" happened, then uses them to discount or at least discredit the possibility. "There is no way one human population could replace everybody else and wipe out their genes, except through violence," he told a reporter from *Discover* magazine. "The people advocating replacement have to come to terms with what they are saying." Violence, as we shall see later, is neither the only possibility nor the most likely one.

Wolpoff and his colleagues do more than simply react against the Out of Africa model, of course. They are enthusiastic proponents of the major alternative, the Multiregional Evolution model. This explanation goes back to Schwalbe's unilinear view of human history at the turn of the century, but more particularly to Weidenreich's version of the model in the mid-1940s and Brace's resurrection of it in the mid-1960s. The definitive statement of the modern version was an 1984 paper by Wolpoff, Wu Xin Zhi, of the Institute of Palaeontology and Paleoanthropology, Peking, and Alan Thorne, of Australian National University, Canberra. In pattern, process, and biology, it is about as different as it could be from the Out of Africa model. Instead of arising in a discrete geographical location, as the Out of Africa model proposes, *Homo sapiens* is said to have emerged throughout the Old World through gradual evolutionary change, the product of existing archaic populations. No significant population movements are called for, nor is there any replacement of populations by evolutionarily advanced people.

The Wolpoff/Wu/Thorne model is, however, not as extreme as certain earlier versions of the unilinear hypothesis that envisioned complete isolation of geographically separate groups (races) stretching back as far as a million years—the model once proposed by Carleton Coon, which implied deep genetic division between living races. The current model argues for a degree of contact among different populations, promoting gene flow between them and effectively main-

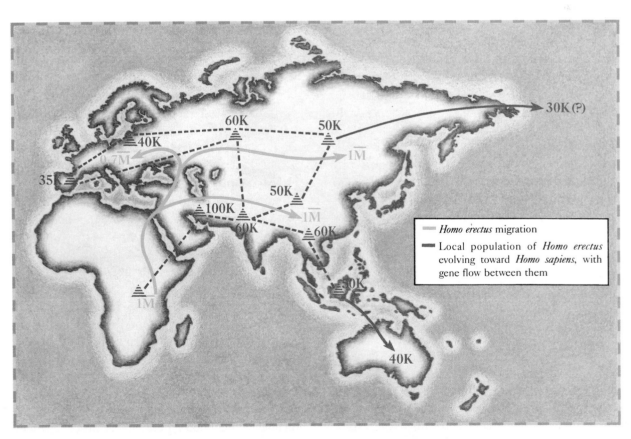

Populations of Homo erectus *moved out of Africa early, and according to the Multiregional Evolution model, evolved locally toward* Homo sapiens, *with a significant gene flow between them.*

taining a network of evolutionary interconnections—a crucial conceptual element. When a species has broad geographic distribution, as had happened by this point in human history, populations in different localities may effectively be isolated from each other by mountain ranges, rivers, or other physical barriers. Gene flow between the populations is then impeded, and genetic differences build up within them.

Such a division is present in certain species in the southeastern United States, where physical barriers (both present and historical) effectively create discrete Atlantic and Gulf coast populations. John Avise and his colleagues at Georgia State University have recently completed a survey of mitochondrial DNA in 19 freshwater, coastal, and marine species; these include the black sea bass, seaside sparrow, American oyster, and horseshoe crab. In virtually all of them the mitochondrial DNA reveals a deep distinction between Atlantic and Gulf coast populations. "It is now abundantly clear that most species should not be viewed as monotypic entities, but rather as a series of geographically differing populations" that have the capacity to breed (and interbreed) successfully, conclude Avise and his colleagues. For a species distributed over several continents, as *Homo erectus* was, barriers to gene flow would be even more severe than revealed in Avise's survey.

Nevertheless, the Wolpoff/Wu/Thorne Multiregional Evolution model argues that although geographically distinct human populations developed some local characteristics, there was sufficient contact among them—specifically, exchange of mates—so that they were able to evolve in concert, imperfect though it may have been. "The Multiregional Evolution model posits a role for both local genetic continuity and gene exchanges," Wolpoff and Thorne explained recently. "Multiregional Evolution begins with the precept that some of the features distinguishing major human groups such as Asians, Australian Aborigines, or Europeans evolved over a long period in the regions pretty much where they are found today." There is an ultimate African source for all these

groups, say Wolpoff and Thorne, but this refers to the population expansion out of Africa of *Homo erectus*, at least a million years ago. Regional differences began to develop at the far reaches of the population expansion throughout Africa, Asia, and Europe, laying down the racial characteristics known today.

The move from *Homo erectus* to archaic *sapiens* to *Homo sapiens* is seen as occurring throughout the Old World with sufficient interaction among populations to maintain a rough contemporaneity. Putting this more graphically, Wolpoff and Thorne cite a description once made by Richard Leakey and the author: "One can think of taking a handful of pebbles and flinging them into a pool of water. Each pebble generates outward-spreading ripples that sooner or later meet oncoming ripples set in motion by other pebbles. The pool represents the Old World with its basic *sapiens* population." In this analogy, the landing place of each pebble represents the point of origin of a unique modern human feature, of which there are many, say Wolpoff and Thorne: "The ripples would denote the spread of the modern features, and where the wavefronts meet each other unique interactive patterns are formed, contributing to the already geographically distinct natures of the populations. With continued pebble throwing, more and more modern features coalesce in this manner, which suggests that there is no specific morphological Rubicon to be crossed to mark the appearance of modern humans."

The process, they claim, is said to be driven throughout the Old World by the effect of the increasing use of technology, releasing anatomical features from biological constraints. Rather than the demands of subsistence shaping the body in the form of powerful muscles and large teeth, tools take the brunt of contact with the environment. As a result, everywhere archaic humans who adopted a greater reliance on technology effectively propelled themselves in the direction of modernity, their anatomy becoming less robust and assembling a constellation of features that characterize *Homo sapiens*. Regional characteristics were maintained through a degree of isolation, while genetic cohesion worldwide was favored by a degree of gene flow. In this model there is no significant mutation that produces a key innovation, such as is implied by the Out of Africa theorists. Instead, Wolpoff and his colleagues see evolutionary change coming about through a shuffling of existing genes among populations over time—with a degree of inexorability, however, about the march to sapienshood, the prime mover being culture.

Disputes over Anatomy

These, then, are the two models that dominate today's explanatory landscape. Different in pattern and process, they make distinctly different projections about what should be found in the fossil record of human prehistory. If the Out of Africa model is correct, three principal predictions should hold: first, anatomically modern humans should appear in one geographical region (Africa) significantly earlier than in others; second, transitional fossils from archaic to modern anatomy should be found only in Africa; third, there should be no regional continuity of anatomical features from ancient to modern populations outside of Africa. In the Multiregional Evolution model the expectations in these three areas are: first, anatomically modern humans will appear throughout the Old World within a broadly similar period; second, transitional fossils from archaic to modern anatomy should be found in all parts of the Old World; third, in each region of the Old World, continuity of anatomy from ancient to modern populations should be apparent.

We have already seen how Stringer's research brought him to the Out of Africa hypothesis, seeing the early appearance of modern *Homo sapiens* in Africa, failing to find transitional fossils outside of Africa, and failing also to see regional continuity of ancient to modern anatomy outside of Africa. The recent establishment of early dates—in the region of 100,000 years—for early modern forms in Africa was an important factor in this equation for Stringer, as for other scholars. "Several years ago I was an adherent of the idea that modern human morphology seemed to have appeared roughly at the same time worldwide," says Fred Smith. "I argued that there was no good evidence for a source area, one region where modern morphology appeared early. Now we have to concede that there is reasonably good evidence that modern human morphology is earlier in Africa than it is in Europe." Nevertheless, Smith sees extensive blending of populations throughout much of the Old World as the early moderns expanded out of Africa. He agrees with Stringer, however, in accepting that population replacement occurred in western Europe, home of the classic Neanderthals.

Smith's position therefore occupies a middle ground between the Out of Africa model and the Multiregional Evolution model, incorporating elements of both. This same position is held by Erik Trinkaus of the University of New Mexico, a prominent scholar on the origin of modern humans. But in today's highly charged and polarized debate, the conceptual middle ground receives little publicity.

Milford Wolpoff's interpretation of the fossil evidence is, of course, different from Stringer's, Smith's, and Trinkaus's. Not only do strict multiregionalists question the interpretation and dating of the putative early modern fossils in Africa, they also point to evidence of unilinear evolution throughout the Old World. "I see continuity everywhere," says Wolpoff, "and even in western Europe I think there was some interbreeding between Neanderthals and modern humans. You see a bunch of Neanderthal features in modern humans." These features range from obvious aspects of anatomy like the shape and size of the nose to less visible details of the skull. For instance, notes Wolpoff, half the Neanderthals had a particular configuration to the opening in the lower jaw into which the mandibular nerve runs. About an equal proportion of early modern people in the region had this same configuration, although it dwindled in

Human Variation

Homo sapiens is widely distributed geographically and has been so for many tens of millennia. As with all such species, local geographical variation has developed at the phenotypic and genotypic levels, but it is remarkably small. The variants, originally geographical, are traditionally known as races and as such constitute a controversial and confused topic in anthropology, usually for social reasons.

The notion that certain races are superior to others—often in unstated realms of achievement and morality—has received explicit support from anthropologists in the past. In 1926, for instance, Henry Fairfield Osborn of the American Museum of Natural History wrote: "The evolution of man is arrested or retrogressive . . . in tropical and sub-tropical regions." Roy Chapman Andrews, also of the American Museum, concurred: "The progress [through evolution] of the different races was unequal. . . . Some developed into masters of the world at an incredible speed," he wrote in 1948.

Explicit racist statements of this kind, common in anthropological texts during the first half of the century, have given way to views in which race itself is often questioned. The earlier notion of "types"—with its implication of pure, isolated groups—has been dropped in favor of continuous variation across populations. Geographical populations may be identified with specific anatomical or physiological features, but these features may also appear in other populations. Population movement through historical and prehistorical times has been largely responsible for blurring the boundaries between geographical groups.

The evolution of geographically characteristic traits is typically considered to be an adaptation to specific prevailing conditions or, less commonly, as a result of genetic drift. Such adaptation has been extensively documented in geographically widespread species of animals and plants, although it is not always possible to link locally evolved features with local conditions. In humans, one obvious variable trait is skin color, assumed to be an adaptation to prevailing intensity of sunlight that reflects the benefits of protection in regions of high UV intensity and the necessity for vitamin D production (which occurs in illuminated skin) in

regions of low UV intensity. Nevertheless, as Harvard University geneticist Richard Lewontin points out, these putative benefits must remain assumptions, as the survival value of dark and light skin at different latitudes has never been tested.

Two very general environmentally induced effects on animal body form were noticed in the nineteenth century. The first, known as Bergmann's rule, is that individuals in colder regions of a species' range have a bulkier body build. The second, Allen's rule, states that in colder regions of a species' range body extremities—limbs, fingers, and ears—will be shorter. Both these effects, clear adaptations to cold, reduce heat loss and are evident in human and non-human populations. Groups of people that have lived for countless generations in hot, arid conditions, such as the Maasai of East Africa, tend to be slightly built with elongated limbs; populations adapted over thousands of years to cold environments, such as Eskimos, are bulky and have short limbs.

Many anatomical traits that may sometimes be thought of being racial characteristics elude easy explanation as adaptations. Texture of hair is one example, as are the presence or absence of the epicanthic fold on the upper eyelid (common in Asiatic peoples), the shape of the lips, and the form of the nipples. Such traits may equally be the consequence of neutral genetic drift, features that have become fixed in geographic populations.

Although anatomical differences between racial groups are undeniable, historically much more contentious is the question of behavioral differences and specifically differences in cognitive abilities. The issue has been muddied for so long by the intrusion of culturally laden judgments and sheer prejudice that no unequivocal answer is possible. Racial differences in cognitive abilities are in principle possible, as a result of adaptation to circumstances that emphasize one kind of ability over another. So far, however, no generally accepted, significant disparities have been demonstrated between geographical races, particularly in intelligence (so-called IQ).

The two prevailing models of modern human origins have different implications for the emergence of racial differences. The Multiregional model states that geographical populations were established as much as a million years ago; significant racial differences might therefore accumulate through time, despite gene flow. There is considerable disagreement among anthropologists over what the fossil evidence implies for ancient ancestry of modern geographical populations. The Out of Africa model, implying recent ancestry of modern populations from descendants of a *Homo sapiens* population that evolved in Africa, suggests shallow racial roots.

Variation among modern populations is considerably less than among archaic *sapiens* populations some 250,000 years ago; moreover, parsimoniously interpreted, this pattern is consistent with modern populations having derived recently from a single, ancestral population.

Recent ancestry of modern humans, and therefore of local geographical populations, is also supported by the basic genetic evidence. The overall extent of genetic variation in mitochondrial DNA among modern human populations is much smaller than among ape populations. Assume that humans and apes diverged by 5 million years ago. If modern human groups were established 1 million years ago, the human mitochondrial DNA variation should be about one fifth that among apes. In fact, the figure is more than one tenth.

Furthermore, as Lewontin established two decades ago, modern geographical populations are extremely similar to one another genetically. Lewontin examined variants in 17 genes, including blood groups and various enzymes, among seven major geographic groups: Caucasians, Black Africans, Mongoloids, South Asian Aborigines, Amerinds, Oceanians, and Australian Aborigines. The result, which was a surprise at the time but has remained robust, is that the great majority of genetic variation—85 percent—occurs *within* racial groups, not among them.

more recent populations. Nevertheless, argues Wolpoff, the point is that it is virtually absent from African fossil jaws of between 100,000 and 200,000 years ago—the putative origin date of modern humans.

"The implications are clear," assert Wolpoff and Alan Thorne. If the early modern populations from Africa replaced archaic populations in Europe, "then the trait must have evolved twice—once among Neanderthals and then again in the European branch of [the African modern's] family tree." A much more reasonable interpretation, say Wolpoff and Thorne, is that "Neanderthals bred with other forms and so made a genetic contribution to later European populations." This example of putative regional continuity of anatomy is one of many, according to Wolpoff and Thorne: "In each region of the world, we have uncovered links that tie living populations to their local antecedents, whose remains are preserved in the fossil record for the area." Of all the areas studied, "the most convincing evidence comes from Asia."

Multiregionalists argue, for example, that in northern Asia various features of the skull—the shape of the face, configuration of the cheekbones, and shovel-shaped incisor teeth—can be traced from

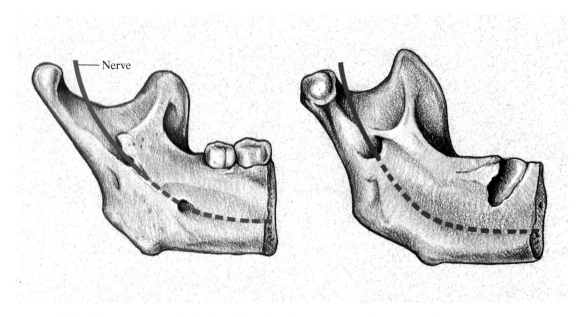

A possible link between Neanderthals and modern European populations may be seen in the jaw anatomy. Supporters of the Multiregional Evolution hypothesis point out that in most living and fossil humans the rim around the mandibular nerve canal opening is grooved (left), while in many Neanderthals it was surrounded by a bony ridge (right). Some later populations also displayed this Neanderthal feature, implying a genetic continuity, albeit limited.

Kow Swamp (about 10,000 years old)

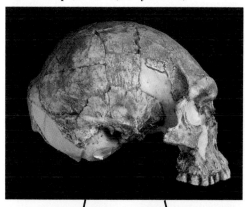

Willandra Lakes
(Upper
Pleistocene)

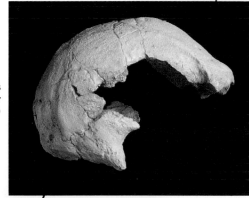

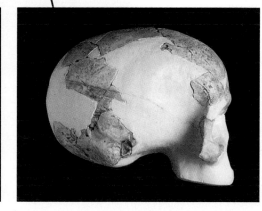

Border Cave
(Upper
Pleistocene)

Sangiran
(Middle
Pleistocene)

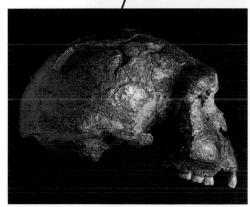

Support for the Multiregional Evolution hypothesis, as offered by its supporters: The progressive changes in the skulls from Australasian sites (Sangiran, Willandra Lakes, and Kow Swamp) suggest the local modern people developed in Australasia over hundreds of thousands of years. Anatomical features of the Border Cave skull, from southern Africa, rule out an African ancestor for the recent Australian population.

fossils 750,000 years old through the famous Peking Man specimens of the Zoukoudian Cave in China to modern Chinese populations. The example most emphasized, however, is from Australasia. "The fossils of Indonesia can be arranged in an anatomical sequence which shows no signs of interruption by African migrants at any time," say Wolpoff and Thorne. "The sequence starts around one million years ago, with the remains of Java Man—a representative of the hominid species *Homo erectus* discovered in 1891—and ends with the remains of Australians dated at around 10,000 years ago." The early Java *Homo erectus* population had large projecting faces, massive, rounded cheekbones, and large teeth—features that can be traced in both overall morphology and in detail through to modern populations, say Wolpoff and Thorne.

All this may well prove to be good evidence for regional continuity of anatomy. But, claims Stringer, many of the anatomical aspects cited are primitive: that is, they appear in many ancient populations and therefore cannot be used as diagnostic clues to particular geographical regions. "It's true that modern Asian populations have some of the highest frequencies of shovelled incisors in the world," he says, by way of example. "And if you look at the Peking Man fossils, they are all shovelled. That looks like good continuity. But when you look at Neanderthals, all those whose incisors are not heavily worn can be seen to be shovel-shaped. You see them in Europe, earlier than the Neanderthals. You see them in *Homo erectus* in Africa. So how characteristic is it of Asian populations? Not at all." Stringer would advance the same line of argument for many of the features said to indicate regional continuity in Asian and other populations.

Australasia, however, is a different matter, not least because the fossil record is so sparse. Although Wolpoff and Thorne speak of "an anatomical sequence which shows no signs of interruption by Afri-

can migrants," the sequence is in fact only spottily populated. It begins with Java Man at about 700,000 years ago, jumps to the Ngandong specimens at about 100,000 years ago, and then on to the Australian fossil record, which begins some 30,000 years ago. This anatomical sequence is more gap than substance; nevertheless, some of the anatomical features in Java Man appear to be echoed in historical Australian populations. One complication is the tremendous anatomical variety that existed among Australian populations between the earliest times there and some 10,000 years ago. Some populations were extremely robust morphologically, possessing thick cranial bone and prominent browridges; others, much less so. Over time the degree of robustness diminished. Was Australia settled twice, one people having robust morphology, the other a more gracile build? Or did the early population simply have an unusually large degree of anatomical variation?

For multiregionalists, the explanation is straightforward: it is derived from an ancient lineage, "the mark of ancient Java," as Wolpoff once put it. Stringer suggests that local evolution of these features is a real possibility, but he admits that as yet he has no complete explanation for the evolutionary history of this geographical region.

Another region that has been particularly important in shaping ideas about the pattern of modern human origins is the Middle East, specifically Israel. From a series of caves on Mount Carmel, a site near Nazareth, and another near the Sea of Galilee, researchers since the 1930s have been recovering many fragmentary fossil individuals, some of whom were buried and all of whom come from this interesting period that encompasses the origin of modern humans. Certain of the fossils look very Neanderthal-like: those from Tabun, Amud, and Kebara. Others, from Skhul and Qafzeh, are distinctly more modern-looking, although relatively robust. Until recently the interpretation of these varied individuals was fully in

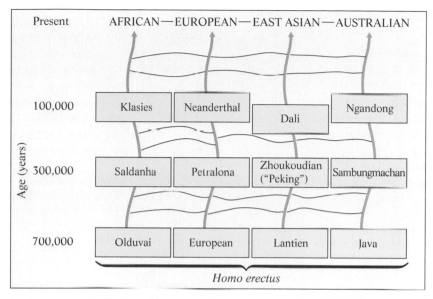

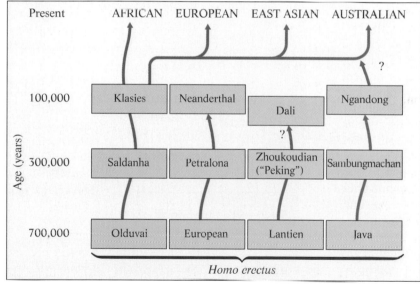

A comparison of evolutionary models: In the Multiregional Evolution model (top), modern regional features are said to have been inherited from ancient populations in the same regions; the unity of all modern populations is said to have been the result of gene flow among contemporary populations (horizontal lines). The Out of Africa model argues that modern regional differences arose recently in populations that had their origin in Africa; existing populations in Europe, Asia, and Australia were replaced by these incoming African populations.

A Revolution in Dating

Anthropologists and archeologists must determine the age of human fossils and artifacts: without such an ability it would be impossible to reconstruct patterns of change in anatomy and behavior through evolutionary time. As we see in this chapter, the recent introduction of two new dating techniques into paleoanthropology revolutionized the interpretation of evolutionary sequences in the Near East concerning modern human origins.

Researchers have two options for seeking the age of fossils or artifacts: direct or indirect methods. Direct methods produce an age for the objects themselves—ultimately the preferred option. There are two types of problem here, however. First, for most material of interest—ancient fossils and most stone tools—there are no methods as yet available for direct dating. Some methods, such as carbon-14 and electron spin resonance, may be applied directly to teeth or young fossils, and indeed to the pigments of rock shelter and cave paintings; thermoluminescence dating may be applied directly to ancient pots or flint tools.

In practice, indirect dating methods represent the typical approach. Here, an age for the fossil or artifact is obtained by dating something that is associated with them. This may involve direct dating on nonhuman fossil teeth that occur in the same stratigraphic layer—by electron spin resonance, for instance. Thermoluminescence can date flint tools associated with human fossils. Ages may

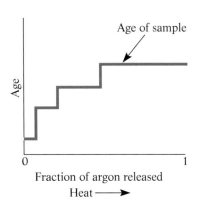

Age of sample

be attributed to fossils or artifacts through information about the evolutionary stage of nonhuman fossils associated with them, a technique known as faunal correlation.

The most common indirect approach, where it is feasible, is to date stratigraphic layers that lie below and above the object in question. Since stratigraphic layers accumu-

The potassium/argon dating technique depends upon the slow radioactive decay of the rare isotope potassium-40 to argon-40, an inert gas. By measuring the accumulation of argon-40 in a mineral sample, an estimate can be made of its age, as long as the absolute amount of potassium is also known. (The ratio of the abundant, stable potassium-39 isotope to the radioactive potassium-40 in natural minerals is constant.) By irradiating the sample with neutrons, the potassium-39 is converted to argon-39. Heating the sample to high temperatures will now release both isotopes of argon, one giving a baseline for potassium content, the other an indication of the sample's age.

A sample may be heated in steps, with increasing temperatures. Argon is released from the outer layers of the mineral grains first. If during its lifetime the mineral suffered no natural heating, the age spectrum obtained by this method will be a horizontal line, giving the age. A sample that has suffered natural heating may have lost some of the accumulated argon-40, giving an artificially young age for the degassed outer layers. Eventually the true age is revealed as a plateau is produced from gas released from the center of the sample.

late from the bottom up, the lower layers are oldest, the upper layers youngest. An object that lies in sediments below a layer dated at 1.2 million years old and above another dated at 1.3 million years can be said to be between these two ages.

Anthropologists these days benefit from a range of dating techniques, which often depend on the steady decay of radioisotopes. The most widely used is the potassium–argon method, based on the decay of the radioisotope potassium-40 to the stable isotope argon-40. Minerals that contain potassium steadily accumulate argon-40, which offers a way of estimating the passage of time. A key factor in all such approaches, however, is that the clock should be set to zero at some point. This occurs in potassium-containing minerals that are ejected during volcanic eruption: the argon-40 in the crystal lattice is released at the high prevailing temperature. Argon-40 found in volcanic material is therefore that formed since the eruption. For this reason the sedimentary layers chosen for indirect dating of fossils are typically layers of volcanic ash.

Potassium–argon dating was first applied to an anthropological question in 1960, when volcanic ash above the famous Zinjanthropus skull found at Olduvai Gorge by Mary Leakey was dated: it showed the skull to be more than 1.75 million years old, rather than the 0.75 million years that had been inferred by other means. A revolution in dating had begun.

The major dating techniques initially available to anthropologists left them with a frustrating problem. Radiocarbon dating could extend to an age of 30,000 to 40,000 years ago, while the reliability of potassium–argon dating declines below about a million years ago. The origin of modern humans falls into this gap. Thermoluminescence and electron spin resonance techniques are able to cover this period, and so when they began to be applied in the mid-1980s there was a potential for another dating revolution.

Both techniques depend on the same phenomenon: the natural irradiation of the crystal structure of the target material, which may be tooth enamel or flint. The source of the irradiation is isotopes of uranium, thorium, and potassium that occur in the soil and within the target material itself. The effect of the irradiation is occasionally to boost electrons from their natural ground state to a higher, excited state. Most fall back to the ground state, but some become trapped at higher states because of imperfections or impurities in the crystal lattice. As time passes excited electrons accumulate in the same way that argon-40 builds up in potassium-containing minerals in volcanic ash. Just as the potassium–argon clock may be reset to zero by heating during a volcanic eruption, so too can the excited-electron clock—again by heating. Materials that have been dated by these techniques include pottery that has been fired and flint artifacts that were fortuitously exposed to heat (a hearth) at the time of use. The crystal lattice of tooth enamel is set at zero as it forms.

Thermoluminescence and electron spin resonance differ in the means by which they detect the quantity of trapped, excited electrons. In the former the target is heated, which causes the electrons to be kicked out of their excited level and fall to the ground state, releasing light: one photon per electron. The amount of light emitted is a measure of how many electrons were trapped at the excited level and of the time of their accumulation.

When electrons are trapped at an excited state they form paramagnetic centers and give rise to characteristic signals, known as electron spin resonance (ESR). The signal is detected by exposing the sample to a variable external magnetic field and microwaves; at a critical strength of the magnetic field microwaves are absorbed, to an extent determined by the number of trapped electrons. The strength of the ESR signal is therefore an indication of the time since the target material was set at zero.

With both techniques the extent of irradiation, internal and external, to which the sample has been exposed must be determined. This is achieved by analyzing the surrounding soil and the sample itself for the relevant radioisotopes. One complication is that uranium tends to leach into tooth mineral as it rests in the ground, thus altering the internal radiation dose. This can be accounted for in calculations, but it may add a margin of uncertainty to electron spin resonance dates. So far bone has proved unsuitable for this dating technique, partly because it absorbs more uranium than teeth, but also because its mineral content is in the range of 40 to 60 percent, compared with 96 percent in tooth enamel. The complexity of changes that go on in the mineral and amorphous phases of the bone during fossilization leads to ESR data that are too uncertain for reliable dating.

Thermoluminescence dating cannot be applied to tooth enamel because chemical changes induced by heat also release energy as light, thus confounding the measurement. Electron spin resonance dating cannot be applied to flint or pottery because the crystal lattices in these materials are insufficiently ordered for the clear production of an ESR signal by trapped, excited electrons.

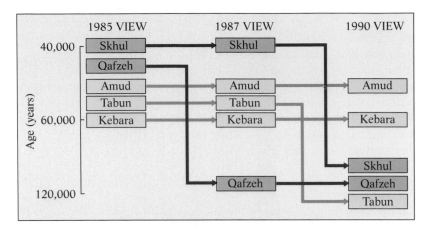

Changing dates changes hypotheses: Until the mid-1980s, Neanderthal populations (in green) in the Middle East were thought to predate, and therefore were ancestral to, modern human populations in the region (left). New dates produced in 1987 (by thermoluminescence and electron spin resonance techniques) placed some modern populations earlier than Neanderthals, thus challenging the supposed ancestral relationship (center). Further dating implied an early contemporaneity of moderns and Neanderthals, which still challenges the original ancestral relationship idea (right).

line with the unilinear model: the Tabun, Amud, and Kebara people were seen as members of populations that eventually evolved into early modern human populations, the people of Skhul and Qafzeh. This picture was shattered in the late 1980s, when new dates for several of the fossil samples appeared to make the neat Neanderthal-to-moderns progression impossible.

Further Evidence

The Neanderthals of the area had been considered to be over 50,000 years old, a position confirmed in 1987 by Hélène Valladas and various colleagues at the Center for Weak Radioactivity, Gif-sur-Yvette, Paris. The researchers employed thermoluminescence

dating to flint artifacts found with the Kebara Neanderthal remains, and obtained an age of close to 60,000 years. The big surprise came early the following year, when the same laboratory produced an age for the modern *sapiens* fossils of Qafzeh: 92,000 years. As Stringer noted in a comment on the published results: "These dates indicate that some modern *Homo sapiens* preceded Neanderthals in the area, standing the conventional evolutionary sequence on its head." If modern humans lived in the Middle East before the Neanderthals, the Neanderthals could not be ancestors of the moderns. The unilinear interpretation falls in the face of this evidence—if, of course, the new dates are correct.

In 1989 more new ages were determined, this time for Skhul (early moderns) and Tabun (Neanderthals) of 100,000 and 120,000 years, respectively. These dates indicate an early presence of Neanderthals in the region, presumably immigrants from Europe. Even though the first modern humans appear some 20,000 years later, an ancestor–descendant relationship between Neanderthals and modern humans is unlikely. The strongest evidence in support of this conclusion is the appearance 60,000 years ago of a Neanderthal specimen at Kebara. If Neanderthals had evolved into modern humans in the region, as the Multiregional Evolution hypothesis holds, then no Neanderthals would be expected after the appearance of modern humans. The Kebara specimen postdates Skhul by 40,000 years.

The Out of Africa interpretation of these fossils and their dates is that Neanderthals moved into the Near East from Europe by 120,000 years ago. Ana-

tomically modern humans migrated into the region by 100,000 years ago. The two populations then coexisted for some 40,000 years. The Kebara specimen's anatomy is classic Neanderthal, showing no indication of interbreeding with anatomically modern humans; in fact, it is one of the most robust and characteristic of Neanderthal skeletons known. Equally, early modern fossils from Israel and Lebanon dated to between 30,000 and 40,000 years ago show no features that might be ascribed to previous hybridization with Neanderthals.

For Stringer this is convincing evidence that Neanderthals and modern humans evolved separately: the former in Europe, the latter probably in Africa. However, if the Skhul and Qafzeh fossils are truly

Kebara Cave in Israel has been an important source of human fossils, particularly Neanderthals.

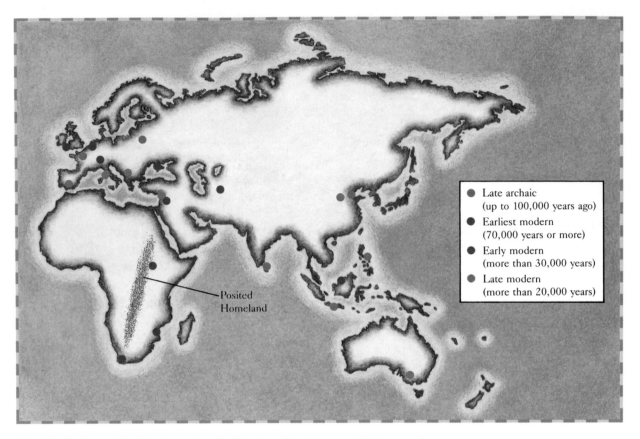

Posited
Homeland

● Late archaic
(up to 100,000 years ago)
● Earliest modern
(70,000 years or more)
● Early modern
(more than 30,000 years)
● Late modern
(more than 20,000 years)

The distribution of fossils shows the distribution of ancient populations as the evolution toward fully modern humans occurred.

close to 100,000 years old, this puts them temporally close to the earliest modern human fossils found in southern Africa. Does this mean that modern humans might have evolved not in sub-Saharan Africa—as the original Out of Africa model supposed—but in East or northern Africa instead? Using currently accepted dates, it does. It also means that the Middle East must be considered as a possible center of origin. Notice, however, how very strongly interpretations of evolutionary pattern depend on accurate dating of fossils.

Multiregionalists, meanwhile, find these interpretations unconvincing. "I've studied all these fossils, and to me they look like individuals from one temporal population," Wolpoff stated at a recent scientific gathering. "Yes, there are some anatomical differences, but my solution is to think of Detroit. If you sampled Detroit, you'd find European anatomy, African anatomy, Middle East anatomy, a whole range of anatomical variation. Just as Detroit has attracted people from all parts of the world, so the Levant was a place to which people were attracted back then; it was a crossroads between Europe and Africa." Wolpoff's statement contains one of the essences of the single species hypothesis: the conviction that tremendous anatomical variation may be accommodated within one lineage, especially over a period of time. Many of Wolpoff's listeners were unconvinced.

As this question—and the historical survey made in our earlier chapters—implies, the interpretation of fossil anatomy lacks a simple, objective measure. "It's often a question of who looks at the fossil material, and what they want to see," comments Stringer. For 140 years, anthropologists looking at the fossils have seen many different things, guided by different intellectual preconceptions. Agreement remains elusive. Under such circumstances, science often benefits from turning to a novel source of evidence that addresses the same issue. This is precisely what happened in mid-1997, with the headline-grabbing announcement of the successful extraction of genetic material (DNA) from the fossilized bones of the original Feldhofer Neanderthal specimen. Researchers in

Germany and the United States were able to compare the structure of this DNA with that in modern people, and found that it differed significantly in terms of nucleotide sequence. The only plausible interpretation of this dramatic new evidence is that Neanderthals could not have been ancestral to modern Europeans and were indeed a separate species, *Homo neanderthalensis*. Arguments by supporters of the multiregional evolution hypothesis for regional continuity in Europe are therefore severely undermined. The genetic data also indicate that modern European populations originated in Africa, as proposed by the Out of Africa hypothesis.

The extraction of DNA from Neanderthal bones has been a dream of anthropologists for many years, because it provides a powerful, independent line of evidence that bears on the debate over the origin of modern humans. It is, however, just the latest (and most spectacular) development in the pursuit of genetic evidence that has been going on since the late 1980s, and includes the famous (some would say infamous) Mitochondrial Eve hypothesis.

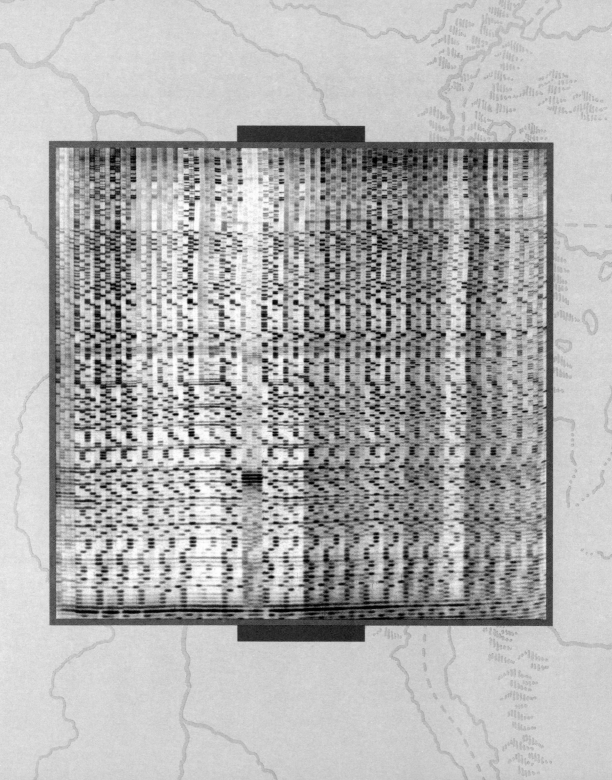

4

MITOCHONDRIAL EVE

In January 1987 Allan Wilson and his colleagues Rebecca Cann and Mark Stoneking of the University of California, Berkeley, published a landmark paper, "Mitochondrial DNA and Human Evolution," in *Nature*. The team of biochemists reported that all living humans can trace part of their genetic inheritance to a single female who lived in Africa some 200,000 years ago. A newspaper article about the work titled "The Mother of Us All—A Scientist's Theory" suggested the biblical appellation Eve, and the name has stuck in scientific and nonscientific writing alike. So-called Mitochondrial Eve immediately assumed an important role in the debate over modern human origins.

For reasons we will soon examine, Wilson, his colleagues, and others had been analyzing DNA from human mitochondria—cellular organelles responsible for generating energy—as a possible way of decoding the evolution of modern humans. The Berkeley group's

Electrophoretic gels, like this one, are used in the analysis of the sequence of DNA and have become an important tool in molecular anthropology, of which Mitochondrial Eve is a prominent part.

Rebecca Cann of the University of Hawaii, seen here examining DNA sequencing gels, was a coauthor on the 1987 Nature *paper that established the idea of Mitochondrial Eve.*

paper, which presented data from 147 human subjects around the world, suggested that the question of modern human origins had been resolved. The mitochondrial DNA (mtDNA) data, noted Wilson and his colleagues, indicated that "the transformation of archaic to modern forms of *Homo sapiens* occurred first in Africa, about 100,000 to 140,000 years ago, and that all present-day humans are descendants of that African population." In other words, it unequivocally supported the Out of Africa hypothesis, soon to be known also as the Mitochondrial Eve hypothesis. In a commentary on the paper published in the same issue of *Nature*, Jim Wainscoat of the Radcliffe Infirmary, Oxford, described the Berkeley study as

providing "the strongest molecular evidence so far in favour of the African population being ancestral [to modern humans]."

Few were aware that the Berkeley group was not the first to tackle the question of modern human origins using mtDNA evidence. In 1983 Douglas Wallace and his colleagues, then at Stanford but now at Emory University, had published data that were essentially the same as Wilson's. One reason why the Berkeley team's results attracted wide attention (which the earlier paper had not) was that their conclusions were boldly stated. Instead of settling the debate, however, the new mtDNA evidence pushed positions to further extremes. Opinions over the validity of the genetic evidence became sharply divided, with anthropologists and molecular biologists among proponents on either side. Wilson was openly dismissive of the contention, asserting that the modern human origins question was "essentially solved" and suggesting that his critics "either don't understand the data or don't take the trouble to look at it carefully." The pitch of the debate sharpened early in 1992 when it was discovered that some of the analysis of the mtDNA data had been inadequate. Opponents of the Mitochondrial Eve hypothesis immediately declared the hypothesis falsified, whereas proponents argued that it remained intact.

Douglas Wallace of Emory University was a pioneer in using mitochondrial DNA in the study of the origin of modern humans.

A Fast-Ticking Clock

Controversy was not a novel experience for Wilson, particularly in connection with human origins research. The fifteen-year battle with the anthropological establishment that he and his Berkeley colleague Vincent Sarich had endured two decades earlier over their proposal for the pattern and timing of hominid origins had generated resentment and suspicion on both sides of the issue. Wilson and Sarich had based their claim on simple immunological measures of differences among albumins, proteins dissolved in blood serum. Contemporary molecular biology tools can analyze genes at the fundamental level of their DNA sequence; the immunological reaction of proteins used by Wilson and Sarich was a crude measure, by comparison. Many differences in DNA sequence do not translate to differences in protein structure that are detectable immunologically, and so the earlier technique was limited in what it could reveal about evolutionary relationships between different proteins. But Wilson's laboratory remained at the forefront of developing new molecular techniques for probing evolutionary histories, including analyzing the DNA sequence of various genes. Comparison of the DNA sequence of a gene from two species represents the ultimate in the ability to read their evolutionary history. Eventually Wilson decided to analyze mtDNA in humans, for two reasons.

First, unlike DNA in the nucleus (the vast majority of the cell's genetic blueprint), which an individual inherits from both parents, mtDNA is inherited only from the maternal gamete. This is because mitochondria lie in the extranuclear protoplasm. When male and female gametes (sperm and ovum) fuse during mammalian fertilization, the sperm contributes only its chromosomes to the newly formed zygote; the ovum contributes its chromosomes and the cytoplasm that surrounds them. Therefore, whether it is destined to be male or female, the developing embryo contains mitochondria only from the cytoplasm of the

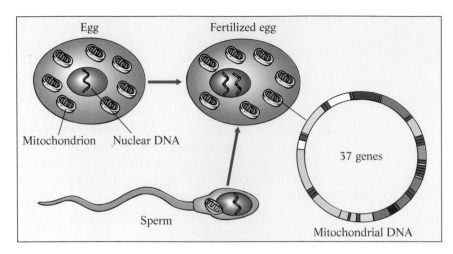

Inheritance of mitochondrial DNA: When a sperm and an ovum fuse, each contributes nuclear DNA in equal amounts. The circular DNA molecule in mitochondria, however, is contributed only by the ovum, resulting in maternal inheritance of this DNA.

ovum—effectively, from the mother. Geneticists who try to trace family trees using nuclear DNA face the confusion generated by the mixing of maternal and paternal genes in each individual. By contrast, because mtDNA is traced back exclusively through mother to maternal

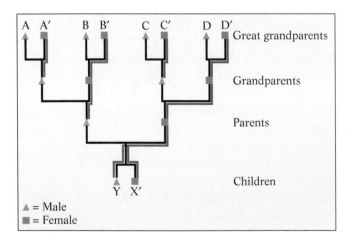

Maternal inheritance of mitochondrial DNA: The male (Y) and female (X) offspring here receive nuclear genes from all eight great-grandparents, but their mitochondrial genome derives from just one great-grandmother, D' here.

grandmother to maternal great-grandmother and so on, it offers more readily interpretable evidence for reconstructing phylogenetic trees.

Second, mtDNA can be a more sensitive molecular clock than nuclear DNA by which to measure the accumulation of genetic differences between individuals. The longer two populations have been separate from each other, the greater the amount of genetic difference that will have accumulated between them. In nuclear DNA, this is a relatively slow process. However, the small, circular molecule of mtDNA accumulates mutations five to ten times faster than the long, linear molecules of nuclear DNA: in the range of one to two base changes in every 100 nucleotides per million years. As a result, mutations in mtDNA can be exploited to record the passage of relatively short periods of evolutionary time—several hundred thousand years.

Underlying all this is the assumption that, on average, mutations accumulate at a regular rate: specific amounts of accumulated mutation can then be related to the passage of specific periods of time. The steady accumulation of mutations is therefore equivalent to the steady ticking of a clock, since genetic difference between species equates to the evolutionary time since they diverged from one another.

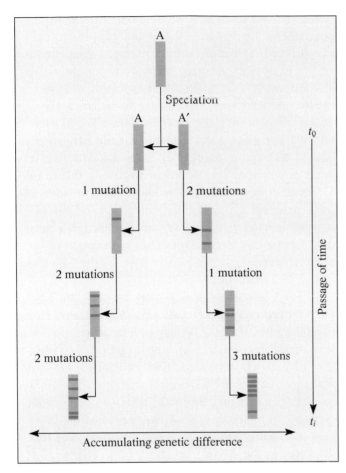

Accumulation of genetic differences: A represents a gene in a species that has undergone speciation (producing two daughter species). Initially gene A is identical in the two species (denoted, however, as A and A'). Mutations then accumulate through time, at a roughly constant rate. In this scheme, gene A accumulates five mutations, gene A' six mutations. The mutation rate is therefore an average of 5.5 per t_1 years; the total divergence rate between the two lineages is 11 per t_1 years.

Predictions Tested

The technique that the Wallace and Wilson groups used initially to analyze mtDNA was restriction fragment mapping, so named because it involves exposing the DNA to restriction enzymes that cut the DNA strand at sites determined by the local sequence. DNA that is treated in this manner generates particular patterns of fragments of various sizes. If the DNA in two individuals is identical, restriction fragment mapping will produce identical patterns. If, however, mutation has altered the sequence at a cut site so that the restriction enzyme no longer recognizes it, or if new cut sites are produced, different patterns of fragment are produced. Because restriction fragment mapping gives information only about the DNA sequence at cut sites, it samples only a limited proportion of the DNA segment being examined. In the experiments performed in Wilson's and Wallace's laboratories, the figure was about 9 percent of the 16,569-nucleotide sequence that constitutes human mitochondrial DNA. (The human mitochondrial genome, the complete sequence of which is now known, is one of the smallest mtDNA molecules and codes for 37 genes. By contrast, the DNA molecules that constitute the human genome—the entire complement of nuclear DNA—extends to 3 billion base pairs and codes for 100,000 genes, only a tiny proportion of which has been sequenced.)

A great quantity of data is produced from restriction fragment mapping of even modest population samples. In principle two kinds of information may be extracted from the populations' mtDNA restriction fragment patterns. First, the extent of mtDNA variation among all geographical populations as a whole, and in comparison with each other, gives an insight into the general history of modern humans; this would include the time of origin of modern humans, and which population, if any, was the longest established. Second, a detailed comparison of the different mtDNA types in the populations may yield a phylogenetic tree that would give a precise description of the evolutionary history of modern human populations.

In the context of the two current models for modern human origins, the mtDNA data would be very different. A multiregional mode of evolution

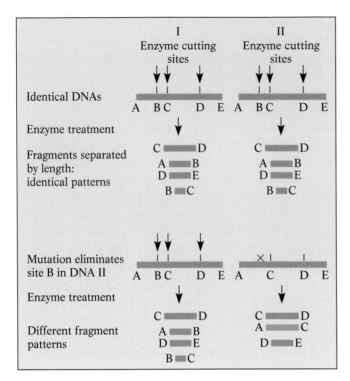

The principle of restriction enzyme mapping of DNA: Restriction enzymes cleave DNA strands at specific sequence sites, yielding fragments of a characteristic size pattern. Individuals with identical DNA will produce identical fragment patterns (top). If a mutation alters a restriction enzyme cutting site the fragment pattern will be altered (bottom).

would produce extensive mtDNA variation in modern populations, reflecting their common origin in *Homo erectus* populations that moved out of Africa at least a million years ago. In addition, all modern populations would display the same degree of diversity. By contrast, an Out of Africa mode of evolution would lead to relatively little mtDNA variation among modern human populations, reflecting a recent origin, while the greatest extent would be present among Africans.

Wilson and Wallace's separate surveys revealed very low mtDNA diversity among modern popula-

tions, just one tenth that seen among African ape populations, for instance. These data appear to indicate a recent origin of modern humans. (Genetic diversity in nuclear genomes of modern populations is also extremely low, again about one tenth that of modern African apes, which may be taken as evidence corroborating the mtDNA data.) The Wilson and Wallace groups also agreed that the African population displayed significantly more diversity than any other population. The Berkeley team took this to indicate an African origin of modern humans, with other geographic populations being descendants of that population. For Wilson and his colleagues these data formed the foundation of a recent origin of modern humans, in Africa: the basis of the Mitochondrial Eve hypothesis.

For Wallace and his co-workers, the data initially looked more equivocal. They noted that if mutations accumulated at a similar rate in all populations, then an African origin was indeed indicated, with a date of 220,000 years ago. They concluded, however, that the greater mtDNA diversity among Africans was the result of a higher mutation rate, not more distant ancestry, in which case Asia was indicated as the region of origin. One reason for this opinion was the discovery in the Asian population of a variant (known as type 8) that also exists in nonhuman primates. The sharing of the type 8 variant with nonhuman primates seemed to indicate that Asians were closer to the ancestral condition and that descendant populations had lost this variant: in other words, Asians appeared to be the longest established of modern human populations.

In their 1983 paper Wallace and his colleagues stated that neither the African origin nor Asian origin could be sustained unequivocally; both interpretations had strong and weak points. A decade later Wallace says that no evidence of different mutation rates has been detected among different geographic populations. One way this was tested was to compare a specific mtDNA sequence in humans from different geographic populations with the equivalent sequence in chimpanzees. If the rate of mutation in one human population was faster than in the others, its sequence would be significantly more divergent from the chimpanzee sequence. In fact, all human sequences tested so far are equidistant from the chimpanzee sequence, indicating an equivalent mutation rate in all human populations. From this and other such tests Wallace concludes that his own diversity data, which now derive from 3065 humans, and Wilson's, which have also been greatly expanded, point strongly to a recent African origin.

The second aspect of mtDNA data analysis, that of reconstructing a phylogenetic tree, is less straightforward and recently has led to great controversy, as we shall see later. The task is to examine what is known of the mtDNA sequence in each individual in the sample and decide how they might be evolutionarily related to each other, based on similarities and differences in sequence. Effectively, one is tracing the path of mutation that took place through the generations from a common ancestor to the diversity seen today. The result is an evolutionary tree of modern human populations. While only one tree actually occurred during prehistory, many potential trees may be reconstructed by analysis of the currently available mtDNA sequence data. The challenge is to find the true phylogeny, or one close to it.

In the early 1980s David Swofford, a systematist at the Illinois Natural History Survey in Champaign, developed a computer program called PAUP, or Phylogenetic Analysis Using Parsimony. The parsimony principle is simple: find the tree with the fewest mutational steps, regarded as the most likely phylogeny. In practice, the task may be daunting. Using data from restriction fragment mapping of the entire mitochondrial genome or, as has happened more recently, sequence information for a fragment of it, the number of possible phylogenetic trees that

may be generated from a sample size of 100 to 200 individuals is enormous: several hundred thousand and perhaps many million. Some of these trees are far more likely than others, but there is no certain way of knowing where in the total universe of possible trees the most likely ones are to be found. Researchers must make a judgment about the number and general structure of trees they wish to sample.

In 1987, reconstructing what they believed to be the simplest possible version of the evolutionary tree from restriction fragment mapping data on 147 individuals, the Berkeley team generated a horseshoe-shaped diagram, now famous, in which there are two distinct branches: one contains only African types, while the second is a mixture of all types. "We infer from the tree of minimum length that Africa is the likely source of the human mitochondrial gene pool," noted the Berkeley researchers. In other words, the phylogenetic analysis using PAUP supported the conclusion based on genetic diversity. The question then was, when did the African mtDNA pool—the first modern humans—evolve?

The answer Wilson and his colleagues produced was between 140,000 and 290,000 years ago. (This conclusion was based on a rate of 2 to 4 percent for the divergence of mitochondrial DNA sequence between two separately evolving groups; it is twice the simple rate of mutation in each of the groups. Wilson had derived his estimate of divergence rate from data on apes, monkeys, horses, rhinoceroses, mice, rats, birds, fishes, and humans. In their tests of divergence rate, Wallace and his colleagues also reached a 2 to 4 percent figure.) The mtDNA data—both from diversity and from the shape of the phylogenetic tree—therefore came down squarely in favor of a relatively recent origin of modern humans in Africa. There is no indication of deep genetic roots (in the form of widely divergent mtDNA types) as predicted by the Multiregional Evolution model.

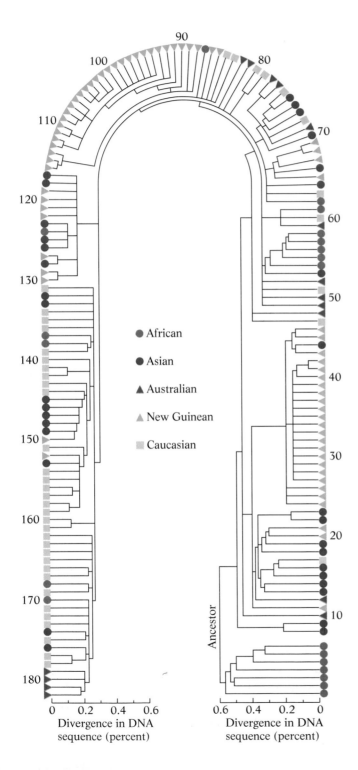

90
100
110
80
70
120
60
130
50

- ● African
- ● Asian
- ▲ Australian
- ▲ New Guinean
- ■ Caucasian

140
40
150
30
160
20
170
10
180

Ancestor

0   0.2   0.4   0.6
Divergence in DNA
sequence (percent)

0.6   0.4   0.2   0
Divergence in DNA
sequence (percent)

Eve's Misleading Mystique

Christopher Stringer, the most prominent supporter of the Out of Africa hypothesis, naturally welcomed the mtDNA evidence. "As has proved to be the case in the study of hominid origins, paleoanthropologists who ignore the increasing wealth of genetic data on human population relationships will do so at their peril," wrote Stringer in *Science* in March 1988, with his Natural History Museum colleague Peter Andrews. It is equally natural that Milford Wolpoff, the multiregional model's most ardent advocate, is the most vocal challenger of Wilson's conclusions. Wolpoff and his colleagues responded to Stringer and Andrews, also in *Science*: "It is appropriate to conclude that paleoanthropologists who ignore the increasing wealth of paleontological data will do so at their peril." The tone of the exchange demonstrates that the old antagonism between anthropology and genetics lingers.

As soon as Mitochondrial Eve made her appearance, the underlying genetic data were vigorously criticized, on several fronts. The rate of accumulation of mutations used by Wilson and others—the ticking of the molecular clock—was said to be too fast. And the interpretation of the "common ancestor" at the root of the tree has also been attacked. "Right from the beginning, there was confusion over what we were saying," lamented Wilson. "And this wasn't confined to the popular press. Some of our scientific colleagues were responsible, too"—as perhaps were

African origin of modern humans based on mitochondrial DNA analysis: Allan Wilson and his colleagues analyzed restriction enzyme maps of individuals from all geographic regions, and produced a pattern of 182 different types (outer edges). The inferred evolutionary relationship among these types implies an African origin. In addition, African populations display the greatest mitochondrial DNA variation, which also supports an African origin.

the Berkeley team's references to a single female who lived somewhere in Africa, some time between 140,000 and 290,000 years ago. (For a variety of technical reasons, this range of dates has since been narrowed; the figure 200,000 years is usually now cited.) Each of us living today, they asserted, derives our mitochondria from this common ancestor: "One lucky mother," Wilson often called her.

Captivated by the powerful image of an African Eve, many people naturally assumed that the Berkeley researchers were talking about a literal Ur-mother—the first modern human, with her Adam. "You have to realize that, in tracing the mitochondrial DNA lineage, we are almost certainly going back to before the origin of modern humans," explained Wilson later. "Mitochondrial Eve was probably a member of an archaic *sapiens* population." More important, the pattern of maternal inheritance of mitochondrial DNA gives only part of the genetic picture, Wilson warned: "She wasn't the literal mother of us all, just the female from whom all our mitochondrial DNA derives." Many of the nuclear genes will have come from other females—and from males. "Eve" was simply one of many individuals in a population from which modern humans eventually evolved.

It is easy to see how confusion arose, and it was virtually assured by a sentence at the end of Wilson's *Nature* paper: "By comparing the nuclear and mitochondrial DNA diversities, it may be possible to find out whether a transient or prolonged bottleneck in population size accompanied the origin of our species."

No Bottleneck Occurred

Biologists speak of a bottleneck when the number of individuals in a population is, for some reason, drastically reduced—by a severe change in habitat, for instance—after which it may once more expand. One consequence of a bottleneck is that the extent of ge-

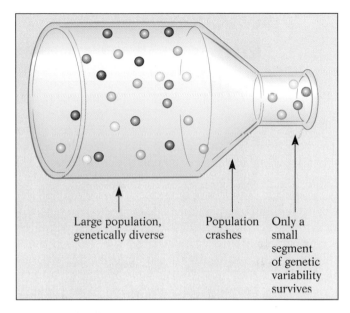

The principle of population bottleneck: A large population will display wide genetic diversity among its members. If the population falls to very low numbers, only a fraction of the existing genetic variation will survive and be passed on to future generations.

netic variation in the new population is greatly reduced compared with the original population: the new population effectively represents only a small sample of the genetic variation that originally existed. Evolutionary biologists consider that bottlenecks may occasionally be associated with the origin of new species; hence the interest in connection with the origin of *Homo sapiens*. The ultimate population bottleneck, of course, would be a single mating couple: Eve and her consort. Jim Wainscoat commented on the Wilson *Nature* paper: "It is tempting to relate the occurrence of the ancestral mitochondrial DNA type to a severe constriction in population size (bottleneck)." Mitochondrial Eve, he seemed to be saying, was a member of a small founding population, if not half of an original pair.

Mitochondrial Divergence: A Dating Strategy

The fact that Africans exhibit the greatest genetic variation of all populations indicates that Africa is the geographical point of modern human origin. Obtaining an accurate date for that origin is crucial. If the origin is recent (that is, in the region of a few hundred thousand years ago), then the Out of Africa model would be supported. However, a date approaching a million years would indicate that the Multiregional Evolution model is correct (the date would mark the migration of *Homo erectus* out of Africa).

The date of the origin is calculated from a measure of sequence evolution that has accumulated in the oldest population and a determination of the rate at which that mutation occurs. Typically, that rate has been calculated by comparing human mitochondrial DNA with chimpanzee DNA (the two species are closely related). Rates obtained this way have led to dates of origin close to 200,000 years ago. There are, however, several potential problems with this comparison.

The first is the assumption that the rate of mitochondrial DNA mutation is the same in the two species. There are many good reasons for thinking that this is the case; nevertheless, it remains an assumption. The second problem derives from the fact that certain sections of the mitochondrial genome, specifically the so-called control region, mutate at substantially higher rates than pertain elsewhere. Within a relatively short period of time, therefore, there is a high probability that mutations in this region will occur at nucleotides that have also mutated earlier. These "double hits" effectively hide the second mutation, as only one base position is changed; further, if the second mutation returns the nucleotide to its original identity, it will appear that no mutation at all has taken place at this position. Molecular biologists have ways of attempting to correct for this process, but obtaining an accurate measure of mutation rate under these circumstances remains difficult. Given the 5 to 8 million years that separate humans and chimpanzees, the double-hit problem is significant.

In an effort to circumvent both these obstacles, Mark Stoneking and his colleagues at Pennsylvania State University adopted a different course. They attempted to calibrate the mutation rate using an *intra*- rather than *inter*specific comparison, using genetic diversity data from human populations in New Guinea. Although it is possible in principle that different populations of the same species might undergo different mutation rates in their mitochondrial DNA, the likelihood is far less than among different species. The human populations in this case lived in the same geographical region and under the same environmental conditions, reducing the chances still further. The problem of obtaining erroneous mutation rates through rate differences is effectively eliminated.

The period of time over which mutation was measured in these populations is 60,000 years, from when the region is considered to have been first colonized by modern humans. This relatively short period obviates the second confounding problem of rate estimation, the double hits.

Stoneking and his colleagues collected mitochondrial DNA data from 50 New Guineans and 20 individuals from Indonesia. These data indicate that New Guinea is populated by three principal groups, descendants of three founding females present in the early settlement. The different

---

The idea of bottlenecks had been very much in the air when Wilson and his colleagues published their 1987 *Nature* paper. In fact Wesley Brown, a member of the Berkeley lab, had raised the idea in June 1980. Reporting on some of the earliest work on human mitochondrial DNA, Brown wrote: "The result indicates that the human species may have passed through a severe population constriction ('bottleneck') relatively recently." Brown, now at the University of Michigan, says that this was merely one in-types of mitochondria in these three original females shared a single ancestor much earlier than the time of migration into New Guinea; their mitochondrial sequences had been diverging from each other before the migration and continued to do so afterward. In addition, the mitochondrial DNA in the descendants of each of the three founding females began to diverge from the original type once settlement had occurred, forming the three major groups of today.

The existence of these groups allowed a measure of mutation rate by two methods. One was to determine the variation that has accumulated within each of the three groups. The second was a measure of the divergence among the three groups. Unless confounded by unexpected circumstances, the two measures should be the same, acting as an internal test of validity. The results were indeed the same: close to 12 percent per million years (for the rapidly mutating control region). This mutation-rate figure leads to an average date of origin of modern humans of 135,000 years, which approaches the age of the earliest known modern human fossils (a little in excess of 100,000 years, in Africa and the Middle East).

Unlike most such calculations, the nature of the data collected here allows the calculation of 95 percent confidence intervals on the average figure given. For the within-group measure, the best estimate for the age of the human mitochondrial DNA ancestor is 133,000 years, with 95 percent confidence intervals of 63,000 to 356,000 years. The corresponding figures for the among-group approach are 137,000 years and 63,000 to 416,000 years. The spread of dates looks large, but this is the nature of their statistical presentation; nevertheless, even if the true date were at the upper end of the spread, it would still preclude the Multiregional Evolution hypothesis, which posits the origin of divergence at 1 million years ago.

Several further observations may be made. First, the mutation rate deduced from the original calculation was based on an assumption that the New Guinea population became established 60,000 years ago. There is in fact very little archeological evidence of occupation earlier than about 45,000 years ago. Currently the evidence for first entry into nearby Australia indicates a date between 50,000 and 55,000 years ago. In employing a date of 60,000 years for New Guinea, Stoneking and his colleagues are being conservative.

Dates younger than 60,000 years would have the effect of boosting the mutation rate and thus reducing the age of the mitochondrial DNA ancestor. Dates older than 60,000 years would have the reverse effect. If occupation occurred 80,000 years ago, for instance, the maximum age of the mitochondrial DNA ancestor (on the 95 percent interval scale) would rise to 464,000 years, still well below the date predicted by the Multiregional Evolution model. As the earliest known modern human fossils in the Old World (in Africa and the Middle East) are just a little in excess of 100,000 years, there seems only a small likelihood that the date of entry into New Guinea would significantly exceed 80,000 years ago.

No doubt controversy and confusion will continue to follow the mitochondrial DNA evidence. And even if all technical doubts were to be dispelled, uncertainty would still remain. The mitochondrial DNA genome is inherited effectively as a single gene, and it is possible in principle that something may have occurred in earlier human populations to distort the true history of this gene, making its common ancestor appear erroneously recent. Evidence on the population variation and history of other genes will be required to test the conclusions based on mitochondrial DNA.

terpretation of his data, but that several other interpretations had been edited out of his paper and the bottleneck idea therefore firmly planted.

Just a year before the Wilson lab's *Nature* paper, moreover, Wainscoat and his Oxford colleagues had published relevant data of a different kind: variants in a nuclear gene region, the beta-globin cluster that codes for several blood proteins, from eight human populations. The variants studied were known as restriction fragment length polymorphisms, variants in DNA sequences at specific sites that may or may not be manifested at the protein level. The variants are detected by the same principle described earlier for restriction fragment mapping in mtDNA. In this case five restriction enzyme cut sites were examined in the beta-globin region. Combination of variants at these sites gives a possible 32 patterns, or haplotypes, of which only 14 occurred in the 600 individuals surveyed.

The pattern of haplotypes among the eight populations in the survey indicated a fundamental split between African and non-African populations, with Africans the longest established. Because of the limited number of haplotypes, Wainscoat and his colleagues further concluded that the origin of the modern human population in Africa had included a bottleneck. "The founder population was small," they stated, and spread from there to the rest of the Old World. An accompanying commentary from London University geneticists Steve Jones and Shahin Rouhani was entitled "How Small Was the Bottleneck?" Their answer was that it could have numbered as few as 40 individuals.

Many observers—molecular biologists and anthropologists alike—concluded that if population bottlenecks could be shown *not* to have occurred in human history, the Mitochondrial Eve hypothesis would be refuted. Such evidence was forthcoming, principally from Jan Klein and his associates at the Max-Planck Institut fur Biologie, Tübingen. Their work concerned genes of the major histocompatibility complex (MHC), which code for a large number of proteins concerned with the body's immunological identity and responses. Their large number and ease of detection at the protein level make the genes valuable markers in population genetics studies.

Klein and his colleagues showed that variants of the MHC genes are shared widely between humans and chimpanzees, indicating that population bottlenecks—which would have reduced the range of MHC variants that humans and African apes have in common—had been unlikely at any point in human history. "The notion of there being a single Eve who was the first 'lucky' mother of us all 200,000 years ago can therefore be discarded," Klein wrote in January 1990.

Klein had first presented his data a year earlier, at a UCLA symposium on molecular evolution organized by Steve O'Brien and Michael Clegg. According to O'Brien, the overwhelming sentiment at the meeting was that the Eve hypothesis was sunk. Klein's data seemed to demonstrate that the human population has never dropped below about 10,000; hence an African Eve—one of a couple or a tiny population—was an impossibility. Reporting on the UCLA meeting in *Trends in Evolution and Ecology*, Craig Packer of the University of Minnesota said: "The garden of Eden, it turns out, was fairly crowded."

But no matter how packed the garden, the Mitochondrial Eve hypothesis is unaffected, because it does not in fact require a population bottleneck. As John Avise had demonstrated in a paper published in 1983, it is possible—indeed inevitable—to go from a population of large size and great diversity of mitochondrial DNA types to a descendant population of similar size that nevertheless carries essentially a single type, all variants of one of the originals. "I was prompted to think about the problem when I saw Wes Brown's 1980 paper on human mitochondrial DNA," recalls Avise. The solution is simple, even if it is counterintuitive at first blush. "It is easiest to think of in terms of family names, an idea Michael Clegg suggested to me."

Assume that the family name is carried by males. Because not all families have male offspring, sometimes a family name will become extinct. In a population of constant size from generation to generation, a pattern quickly develops. In the first generation, on average, one quarter of the couples are expected to have two boys, who will each carry the family name to the next generation; one half will have a boy and a girl, the family name being carried by the single male child; and one quarter will have two girls, the family name disappearing. The same

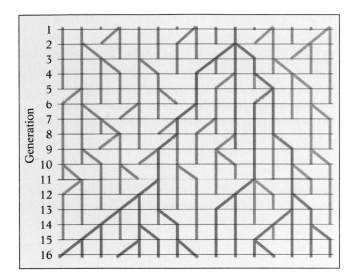

The role of change in maternal mitochondrial DNA ancestry: Mitochondrial DNA passes only through the maternal line. A mitochondrial DNA type in a mother who produces only sons will therefore become extinct. With the passage of generations all but one mitochondrial DNA type will become extinct for purely stochastic reasons, leaving a single ancestral type.

process continues through the generations, so that after a certain period of time (about twice as many generations as there are females in the first generation), only one family name remains. Thus a great diversity of names has been reduced to just one, simply by stochastic loss. The same holds true for mitochondria, except that transmission is through the female line.

A large population, then, given sufficient time, will steadily shed mitochondrial lineages as a stochastic process, eventually finishing up with one shared by all individuals. Although they played a prominent role in the unfolding discussion of the Mitochondrial Eve hypothesis, population bottlenecks turned out to be a red herring in this connection, and the demonstration of their absence in human history is of no relevance to the hypothesis.

Implications of Complete Replacement

In the first experimental data set that Wilson and his colleagues published there were just 147 individuals represented—a relatively small but significant number—and in all of these the mitochondrial DNA was relatively young, derived from a single type in an African population some 200,000 years old. If the descendants of that population had bred with existing populations of archaic *sapiens* as they expanded into the Old World, ancient mitochondrial DNAs (from lineages stretching back to *Homo erectus*) would have been incorporated into the gene pool of modern humans—in which case, individuals with these ancient mitochondrial DNAs would be expected at some point to turn up in an experimental sample taken today. None was present in the initial sample of 147. Since then Wilson, his colleagues, and workers in other laboratories have sampled more than 4000 non-African individuals, still failing to detect ancient DNA. "Perhaps we've been unlucky," observed Wilson wryly, "but I don't think so. By now I'm pretty sure that we haven't found any because none exists."

In September 1991, two months after Wilson died, his name appeared on a paper in *Science* together with those of fellow molecular evolutionists Linda Vigilant and Mark Stoneking and anthropologists Henry Harpending and Kristen Hawkes. The paper presented the latest effort to test the Eve hypothesis, examining the sequence of a rapidly evolving part of the mtDNA (the control region) in 189 individuals. The new study included native Africans and incorporated comparative sequence data from a chimpanzee, man's closest genetic relative. The results, consistent with those of the earlier work, were bolstered by two powerful statistical tests. A phylogenetic tree was constructed using PAUP, after testing 100 possible trees. "Our study provides the strongest support yet for the placement of our common mitochondrial DNA ancestor in Africa some 200,000 years ago," the authors concluded.

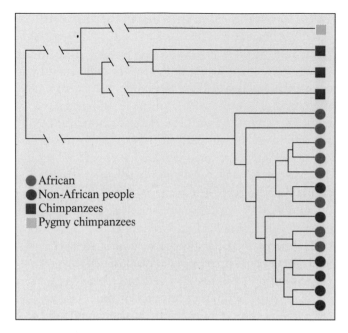

● African
● Non-African people
■ Chimpanzees
■ Pygmy chimpanzees

A method for estimating modern human divergence: The diagram shows accumulated divergence of mitochondrial DNA in human and chimpanzee populations. Divergence among humans is less than one twenty-fifth of that between humans and chimpanzees. Vincent Sarich estimates the human/chimpanzee divergence to be close to 5 million years ago, giving a date of less than 200,000 years ago for the divergence of modern human lineages.

The implication of these results—specifically, of the absence so far of any known ancient mitochondrial DNA in modern people—is that when modern humans expanded out of Africa, they failed to interbreed with established populations; rather, they eventually replaced them. Complete population replacement, we recall, was not a requirement of Stringer's Out of Africa model, which had allowed for some population interbreeding. "It's a tough thing to contemplate, complete replacement," says Stringer. "But the logic of the mitochondrial data seems strong. It is difficult for us to be as unequivocal with the fossil data."

He does point out, however, that the sparsity of the fossil record in most regions of the world precludes a confident conclusion. "It may be significant that in the area where we do have the best fossil record, western Europe, we are able to conclude that complete replacement took place—of the Neanderthals," he adds. "Maybe we would be able to make the same conclusion if we had a better fossil record in other areas."

Potential Problems with DNA Analysis

Wolpoff and others remain unimpressed with the conclusions from the Berkeley laboratory. There are several aspects in which the mtDNA in today's human populations is artificially young, Wolpoff suggests: both concern ways in which the loss of mitochondrial lineages might be distorted, thus giving an erroneous picture of population history.

Initially, however, Wolpoff's concern was about the accuracy of the mitochondrial DNA clock, specifically its mutation rate. At the annual meeting of the AAAS in February 1990, Wolpoff suggested that Wilson and his colleagues had misread the ticking of the mitochondrial molecular clock, estimating a rate of accumulation of mutations too high by as much as a factor of four. The figure should be 0.715 percent, he said, not 2 to 4 percent. "That would make Eve about 800,000 years old, which is just fine by me," said Wolpoff. In fact, 0.715 percent, the product of calculations of Masatoshi Nei of Pennsylvania State University, was an inappropriate comparison with Wilson's 2 to 4 percent rate, an error that has been made repeatedly by critics of Mitochondrial Eve. The 2 to 4 percent rate reflects the *divergence between two lineages*, and so counts mutations in both lineages. Nei's figure of 0.715 percent is a *substitution rate in one lineage*. The proper comparison is therefore between 2 to 4 percent (Wilson and colleagues) and 1.43 percent (Nei).

This still leaves a difference of some 200,000 years (Wilson's calculation) as against 400,000 years (derived from Nei's data) as the putative origin of modern humans. But 400,000 years still does not take Eve back to the beginning of the *Homo erectus* settlement of the Old World at least a million years ago, as multiregionalists argue.

Since the molecular clock is at the heart of the Mitochondrial Eve hypothesis, the question of how fast it ticks is clearly vital. According to Wilson, Nei is too cautious in his analysis and hasn't carefully studied the Berkeley group's calculations. In response to this, Nei says that Wilson has not taken account of factors that may artificially increase apparent mutation rate. No resolution of these differences is yet in sight, but for many researchers in the field, including Wallace's group, the 2 to 4 percent range looks acceptable.

By the spring of 1990 Wolpoff had begun to argue that the clock is irrelevant. With heavy irony he recently conceded that "the Wilson lab got the rate precisely right." Nevertheless, he now says that "with stochastic loss of mitochondrial lineages, it's impossible to calculate any rate reliably." A population would finish up with an artificially young array of mitochondrial lineages if older lineages were lost preferentially, notes Wolpoff, and unusual population demographic histories might promote such a distorting effect. Mark Stoneking rebuts this suggestion by emphasizing that the loss of mtDNA is stochastic, and so offers no reason why older lineages might preferentially be lost. Stochastic loss should produce a true reflection of the population's overall history in the lineages left behind. If the loss is severely biased, Wolpoff's argument would have merit; no examples of consistent bias of this sort are known, however, among animal populations studied so far.

Wolpoff's second argument is related: if one mitochondrial type has even a slight advantage, it will be favored and will spread more rapidly through the population. "This would eliminate other types, and make it look as if the mitochondrial DNA population as a whole were young," says Wolpoff. John Avise agrees that, in principle, selection could distort the picture. "A mutant could sweep through the population, it's true," he says. "You'd have to have strong contact across the entire population for it to happen, but it's always a worry in basing phylogeny on mito-

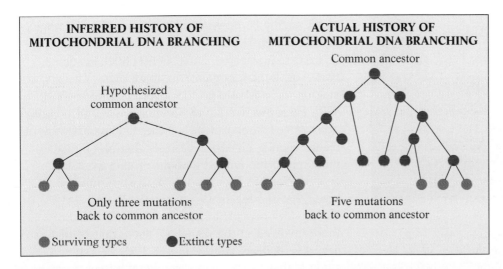

INFERRED HISTORY OF MITOCHONDRIAL DNA BRANCHING

Hypothesized common ancestor

Only three mutations back to common ancestor

ACTUAL HISTORY OF MITOCHONDRIAL DNA BRANCHING

Common ancestor

Five mutations back to common ancestor

● Surviving types   ● Extinct types

Supporters of the Multiregional Evolution hypothesis argue that maternal lineage reconstructions based on surviving mitochondrial DNA types are inherently flawed. A bias toward the loss of older lineages would give an erroneously young ancestral date, as shown. However, no such bias has been shown to occur in the stochastic loss of lineages in animal populations so far studied.

chondrial DNA analysis." So far no examples of such a process have been detected with any species, but it remains a theoretical possibility.

A Premature Obituary?

In February 1992 the pitch of debate over the validity of the Mitochondrial Eve hypothesis increased markedly. In that month two short papers were published in *Science*, one by geneticist Alan Templeton of Washington University, the other by Mark Stoneking and three colleagues at Pennsylvania State University. Both indicated that Wilson's team had inadequately applied the PAUP analysis to their mtDNA data. For their September 1991 paper, Wilson and his colleagues had examined 100 possible trees and selected

Mark Stoneking of Pennsylvania State University, a former colleague of Allan Wilson, is a strong proponent of the utility of mitochondrial DNA evidence in the study of the history of human populations.

one, with an African origin, as the most parsimonious. Templeton reported that he had examined a small number of possible trees and had quickly found 100 that were two steps more parsimonious than Wilson's. Moreover, an Asian—not an African—origin seemed to be indicated by these trees. Stoneking and his colleagues had been alerted to the problems with their parsimony analysis and had generated some 50,000 trees from the mtDNA sequence data, each of which was more parismonious than any of Templeton's. No single geographic origin was strongly supported: Asia was as likely as Africa.

Others were involved directly in the reanalysis of the mtDNA data, specifically David Maddison and Maryellen Ruvolo of Harvard University and David Swofford. Late in 1991 Maddison had reported his analysis of 10,000 PAUP trees based on the mtDNA sequence data, saying that no single geographic origin was indicated unequivocally. Maddison, extending these analyses with Ruvolo and Swofford early in 1992, reiterated the conclusion that PAUP analysis was unable to locate a geographic origin of modern humans. The reason that PAUP does not produce an unequivocal answer stems from the nature of the data. If each geographic population displayed a highly characteristic set of variants (with the ancestral population showing links to all), then an unequivocal tree could readily be determined with the limited amount of data available from restriction fragment mapping or from sequencing a small section of the genome. However, human populations share many variants, making their genetic profiles overlap considerably. With extensive overlap of variants among all populations comes an enormous number of possible ways of reconstructing phylogenetic trees that accommodate them, given the limited amount of data available.

Critics of Mitochondrial Eve were quick to argue that the revelations about the flawed PAUP analysis destroyed the hypothesis. While some of those involved in the reanalysis agree that Mitochondrial Eve is dead, others do not. "All we can conclude is that PAUP cannot resolve the issue with the kind of data we currently have," says Ruvolo. "This is far from an obituary for Eve." The issue might be settled with new data, she says; specifically, complete sequences of between 10 and 20 mitochondrial genomes from individuals in the major geographic populations might be sufficient to resolve the general shape of the tree. Even given current automated DNA sequencing technology, such a goal, though feasible, represents an enormous task.

With PAUP analysis on hold, the details of the phylogenetic tree of modern humans remain to be established. The original horseshoe-shaped tree must be regarded as wrong in detail, and perhaps also in being rooted in Africa. Nevertheless, the mtDNA diversity data and the conclusions drawn from them remain unchanged. For many, including those associated with the Wilson lab, Douglas Wallace, and Maryellen Ruvolo, these data are sufficient for them to paraphrase Mark Twain and suggest that reports of Eve's death have been greatly exaggerated.

Corroborative Evidence

For those who accept mitochondrial DNA evidence as a reliable indicator of evolutionary history, the data collected by Wilson and others look very persuasive indeed, lending powerful support to the Out of Africa hypothesis. For those who worry about potential distortion through various demographic and other influences, or that in the PAUP reanalysis lies a fatal trap for the hypothesis, the conclusions may always remain suspect. In any case, as in all branches of science, other lines of evidence should be sought for corroboration. Mitochondrial DNA in effect acts like a single gene, so confirming data might be sought in other genes. In the end there should be a pattern of concordance between mitochondrial and nuclear genes.

As noted earlier, data on nuclear genes must be extracted from a very "noisy" background of maternal and paternal contributions—a technical challenge.

Nevertheless useful data are accumulating, of two main forms. The first are the "classic" genetic markers, such as blood groups, and variants of certain immunological and other proteins, about a hundred of which have now been determined in thousands of human populations. The second are restriction fragment length polymorphisms, several thousand of which have been discovered in recent years, but surveyed in only a relatively few populations. The laboratories of Luigi Luca Cavalli-Sforza at Stanford University and of Masatoshi Nei at Pennsylvania State University have been at the forefront of collecting these two kinds of data in human populations.

Preliminary assessment of classic genetic markers collected by the mid-1960s appeared to indicate to Cavalli-Sforza an Asian origin of modern humans. As more data became available, however, he became persuaded that an African origin was more likely, a position he first stated strongly at a Cold Spring Harbor meeting on the "Molecular Biology of *Homo sapiens*" in May 1986. Data on 44 classic genetic markers in 42 populations and 80 DNA polymorphisms in 8 populations indicate an initial split between African and non-African populations, he says, which is highly suggestive of an African origin. The classic genetic marker data delineate three subsequent splits: first, between southeast Asians and Pacific Islanders from northeast Asians, Caucasoids, and Amerindians, which suggests two migrations into Asia; next, Caucasoid from northeast Asians and Amerindians; and last, Amerindians from northeast Asians. The first two of these are also seen in the DNA polymorphism data (Amerindian data were not included there).

It is not yet possible to place a time scale on these divergences directly from the DNA data. Nevertheless, the genetic distances seen in the tree constructed from the classic genetic marker data coincide in relative terms with dates for the four splits as determined from archeological evidence. The consistency between archeological and genetic data "lends support to the hypothesis that the genetic tree corresponds roughly to the evolution of human populations over the last 100,000 years," Cavalli-Sforza told a scientific meeting at the Royal Society, London, at the beginning of 1992. Masatoshi Nei reaches the same general conclusion after surveying classic and DNA polymorphism markers. "Genetic data seems to be in favor of a single origin hypothesis," he stated recently.

In 1988 Cavalli-Sforza extended his comparison of genetic and archeological data to include linguistic data and found an impressively close match. Languages, like species, evolve over time, and it is possible in principle to reconstruct a linguistic phylogeny. Cavalli-Sforza thought that the phylogeny might illuminate the population history of modern humans.

Ancient demographic movements inevitably impress themselves in various ways on the prehistoric record. Many of these would involve aspects of material culture, but the spread of language would be indicative too. If the multiregional model were correct, then languages in different parts of the Old World would have extremely ancient and disparate roots, as with genetic heritage. The likelihood of relating patterns among them and their local populations would be very small. If, however, the Out of Africa model were correct, then it is possible that a single language was associated with the founding population of modern *sapiens*—a language that spread out with migrating groups, evolving locally as it went, just as genes do.

It was a fortuitous coincidence that allowed Cavalli-Sforza recently to test this idea. One of this country's most eminent linguists, Joseph Greenberg, is also at Stanford, and that propinquity inspired

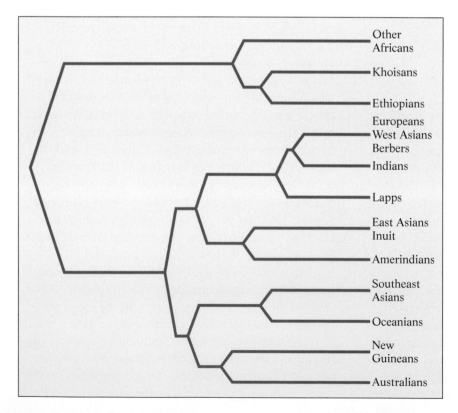

Other Africans

Khoisans

Ethiopians

Europeans
West Asians
Berbers

Indians

Lapps

East Asians
Inuit

Amerindians

Southeast Asians

Oceanians

New Guineans

Australians

Evidence from nuclear DNA supports an African origin. Here, data from the laboratory of Luigi L. Cavalli-Sforza of Stanford University separate African and non-African populations and indicate an African origin.

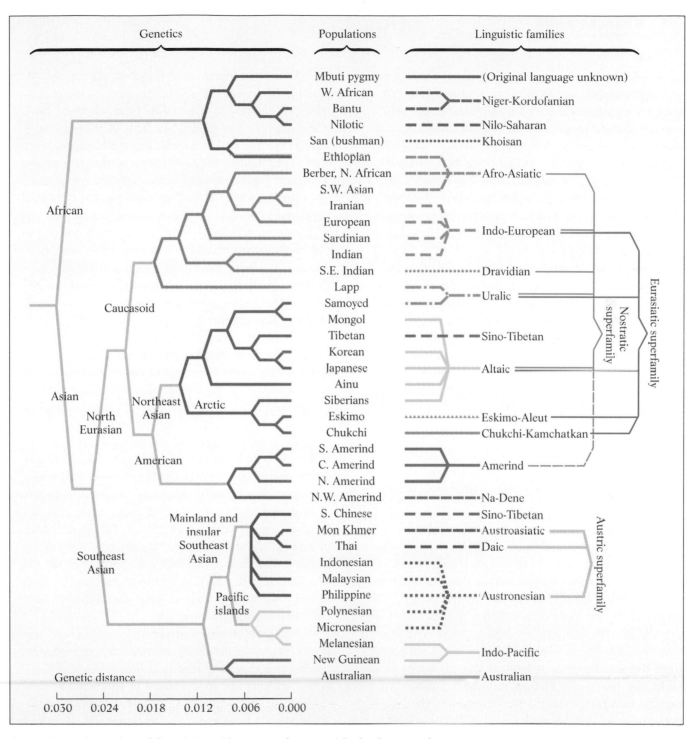

Comparison of genetic and linguistic evidence reveals a surprisingly close match among world populations. The match further encourages the notion of a relatively recent origin of modern humans, diverging from an ancestral African population.

Cavalli-Sforza to test how his extensive genetic data and the archeological data of the Old World might match with linguistic data for the same region. He already knew that the genetic profiles of human populations closely coincided with patterns in the archeological record. The task was to determine whether specific languages or language families mapped precisely with modern populations as defined by their genetic profiles. The match was remarkably close, says Cavalli-Sforza, "indicating considerable parallelism between genetic and linguistic evolution."

Cavalli-Sforza and Greenberg believe that the correspondence between these sets of data is highly significant historically, but many linguists argue that the interpretation is too simplistic. If the Stanford researchers are correct, however, their data support the notion that language was among the final adaptations leading to modern humans; futhermore, they are consistent with a discrete African origin of modern humans.

Predictions from Population Genetics

It is fair to say, therefore, that genetic evidence—from mitochondrial and nuclear DNA—support the Out of Africa hypothesis. Indeed, molecular data pushed the hypothesis even further than its anthropological authors initially envisaged, by indicating complete replacement of established archaic *sapiens* populations by incoming modern humans. The Multiregional Evolution hypothesis receives no support from the genetic data, at least as interpreted by most molecular biologists. The addition of linguistic data extends that position. It is possible to test the two hypotheses further by applying such aspects of biology as population genetics and evolutionary ecology. In the process, modern human origin becomes one of the most thoroughly biologically based issues in anthropology.

With hypotheses as different as the Out of Africa and Multiregional Evolution models, many contrasts are implied in their supporters' views of evolutionary tempo and mode. Not least of these is the question of the extent to which our ancestors' possession of technology might modify evolutionary change. The Out of Africa model makes no case for technology creating a special evolutionary pattern. The Multiregional Evolution model, however, invokes culture as a important agent—one that has produced a special evolutionary pattern not seen in other animal groups. Whatever the merits of these different approaches, both must try to address two basic correlates of the origin of modern human populations: anatomy and genetics.

The origin of modern humans, as we saw in Chapter 1, involved a considerable reduction in the robusticity of the skeleton, both in the cranium and the rest of the body. This implies that modern humans routinely experienced a much-reduced amount of physical stress, compared with their archaic forebears. Was technology now achieving by skill what had previously been a matter of brute force? A second aspect of anatomical change is apparently less significant functionally, but more informative in terms of evolutionary pattern. Among archaic populations, a great degree of variation existed on the basic theme of archaic *sapiens*. With the appearance of modern humans, anatomical variability dropped dramatically. The Out of Africa model is able to explain this readily, on a theoretical level at least; the Multiregional Evolution model, much less so.

The great anatomical variability of archaic populations is clearly the result of their having been established for a very long time—as much as a million years—in separate geographical populations. Populations of the same species in different locations tend to drift apart genetically and anatomically, even with some exchange of genes among the populations. If modern humans evolved from just one of these populations (the Out of Africa model), replacing all existing archaic people as they populated the Old World, then overall variability among modern humans will inevitably be much lower than existed among the archaic populations. In trying to explain this same observation, the Multiregional Evolution model stretches population genetics theory rather more, since it requires that geographically separate and anatomically disparate populations anatomically converge onto the much less variable modern human anatomy. Such a pattern is possible, but it demands an extent of parallel evolution that makes most evolutionary biologists uncomfortable.

The second aspect of modern human populations that must be accounted for by a successful origins model is the surprising lack of genetic variation among modern human populations. Human populations differ genetically among themselves only one-tenth of the degree that would be expected by comparison with most large primate species, for instance. Most genetic variation among modern humans as a whole is accounted for by variation among individuals, not between populations. Such a pattern would be readily consistent with the Out of Africa model, in which all modern humans derive recently from a single genetic pool. The Multiregional Evolution model must argue that genetic homogeneity among populations has been achieved by extensive gene flow among them, a development about which Wolpoff and his colleagues are somewhat ambiguous, often describing gene flow in the model as "enough but not too much." And Wolpoff has suggested that "it is probably the movement of ideas that is critical to the origin of modern humans, more so than the movement of genes."

In the late 1980s, population geneticists became involved in the problem of modern human origins. Specifically, some geneticists asked: What do theoretical considerations say about the feasibility of the two models, particularly over the issue of gene flow? Populations change genetically through a two-step process: first a mutation arises in one location (one individual); second, the modified gene becomes established in the population as descendants of that individual mate with other members of the population. It

is a relatively slow process and requires that there be unbroken chains of mating opportunity throughout the population. The new gene spreads as a wave front from its point of origin and eventually is said to reach equilibrium when the gene is equally common at the point of origin and in parts distant from it.

The two models are very different in their requirements for gene flow. The Out of Africa model calls for a speciation event in a geographically restricted (though not necessarily small) population, from which descendants move into other regions of the Old World. Gene flow, or interbreeding, occurs within the founding population and then between neighboring groups of descendants. There is no requirement for extensive gene flow over distant geographical regions in the process of evolution of modern humans. By contrast, the Multiregional Evolution model requires extensive gene flow over large geographical regions and through long tracts of time; the disparate ancestral populations must be linked genetically as they evolve more or less in concert to the modern condition.

According to most prominent authorities in population genetics, the multiregional model finds little or no support. For instance, Shahin Rouhani concludes that it is theoretically implausible because the extent of gene flow required is both too large and too geographically extensive. "Even under ecologically identical conditions, which is rarely the case in nature, geographically isolated populations will diverge away from each other and eventually become reproductively isolated. . . . It is highly improbable that evolution would take identical paths in this multidimensional landscape," says Rouhani.

Under optimum circumstances, the spread of an advantageous mutation might take half a million years to travel from South Africa to the coast of China, calculates Rouhani. Favorable conditions rarely obtain, moreover, and mating between individuals in neighboring populations is often impeded by cultural and geographical barriers. This effectively rules out the high degree of genetic continuity

Genetics, Linguistics, and Archeology

When one line of evidence points in a particular direction, it may be strong enough to swing a scientific argument. When three independent lines coincide, the argument they support must be accepted as cogent. Such a coincidence of evidence has recently occurred over the origin of modern humans: it comes from genetics, linguistics, and archeology. The strongest of the three data sets is the genetic evidence.

For four decades the Stanford geneticist Luigi Luca Cavalli-Sforza has been collecting genetic data with the aim of building an atlas of human variation. This ambitious compilation of data would allow the attainment of a major goal: "the reconstruction of where human populations originated and the paths by which they spread throughout the world." That goal, says Cavalli-Sforza, is now at hand.

The data that Cavalli-Sforza and his colleagues have accumulated are known as classic genetic markers. All such markers are variants of proteins and include the well-known blood groups A, B, O, and the rhesus (Rh) blood factor. The researchers now have data available on more than 100 genetic markers from 3000 samples taken from 1800 aboriginal populations around the world. Most of the samples include hundreds and in some cases thousands of individuals. As can be appreciated, the power of genetic analysis comes not only from the nature of the data themselves but also from their enormous quantity.

Over the millennia there has been genetic mixing among human populations, particularly where major migrations have occurred. Nevertheless, a strong degree of genetic cohesion has been maintained in local populations, as evidenced in the identity of ethnic or racial groups worldwide. The genetic approach seeks to uncover the genetic relatedness among these groups, effectively leading to a reconstruction of a phylogenetic tree. The technique is based on the fact that populations separated from each other will accumulate genetic differences, simply by drift. For the classic genetic markers, genetic distance between two populations is represented not by a presence or absence of the trait but by its relative frequency in populations.

For instance, the rhesus blood factor has two forms (alleles), Rh-negative and Rh-positive. The Rh-negative allele is common in Europe, infrequent in Africa and West Asia, and virtually absent in East Asia and among aboriginals of America and Australia. If the rate at which the frequency of this allele drifts in the different populations is constant through time, then the genetic distance data serve as a clock, indicating how long each population has been separate from the others. Interpretations based on one gene would be tenuous, but based on a hundred genes, they become powerful.

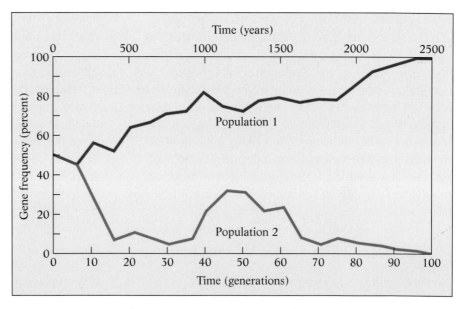

Two populations can drift apart by chance changes. Here a computer simulation shows the outcome of the separation of a population into two parts, beginning with identical gene frequencies. Chance changes in gene frequencies generate substantial genetic difference over time.

The principal conclusion that may be drawn from the body of genetic information now available to Cavalli-Sforza and his colleagues is that modern human populations have an African origin. This may be taken as referring to the migration of *Homo erectus* from Africa to the rest of the Old World more than a million years ago; or to a relatively recent evolution of modern humans in Africa. There is no way to place an absolute time on the genetic distance data as they stand; however, it is possible to look at relative genetic distances and see how they match with predictions of the models and data.

The genetic data identify seven major ethnic populations: African; Caucasian; Northeast Asian; Amerindian; and Southeast Asian in the Pacific Islands, in Australia, and in New Guinea. They show that, given the genetic distance between Africans and non-Africans as 1.0, that for Australians and Asians comes to 0.62, and for Europeans and Asians is 0.42. These genetic distance ratios fit relatively closely with figures taken from one reading of the fossil record: 100,000 years for the separation between Africans and Asians, about 50,000 for that between Asians and Australians, and 40,000 years for that between Asians and Europeans. This matches the Out of Africa interpretation for that first separation. If a figure of 1 million years is imposed on that first separation, as proposed by the Multiregional Evolution model, the ratios fail for the later separations, particularly for the establishment of Australian populations, which would be put at an improbable 600,000 years ago.

The coincidence of genetic and archeological evidence is joined by linguistic evidence. Languages, like genes, change through time and retain a coherence with ethnic groups. Until recently some 5000 languages existed. If modern humans evolved recently, as the Out of Africa model proposes, these 5000 languages must be the descendants of a single, original language, which some call the Mother Tongue. The Multiregional Evolution model expects no single origin of language, however, and modern tongues may be the descendants of a multiplicity of original languages historically unconnected to each other. An analysis of the historical relationships among existing languages should, in principle, resolve this issue—but the rate of change of languages is so high that attempts to trace phylogenies back more than a few thousand years are unfeasible.

Notwithstanding the enormous obstacles, several linguists have attempted this task, with some success. By looking for fundamental, shared linguistic features, some 17 language superfamilies, or phyla, have been constructed; examples are Indo-European, Amerindian, Indo-Pacific, and Altaic. Cavalli-Sforza and his colleagues compared the language phyla with the genetic groups they had identified and found a very close correlation: in almost all cases a language phylum coincided with ethnic populations linked genetically. Anthropologists have been aware of the close correlation between tribal identity and language, but the extent of correlation revealed by this work, linked as it is with genetic data, is surprising: it implies a cohesion that goes far back in prehistory.

Although no linguist claims to be able to reconstruct the language phylogeny back to its origin, some argue that historical links can be discerned among the phyla, forming superphyla. These superphyla, if valid, are looking back at least 10,000 years—perhaps twice that far—into linguistic history. One such superphylum, known as Nostratic, links languages spoken by Caucasoids and northeast Asians. A further step back into prehistory may link Nostratic with the Amerindian language phylum. Again, there are close genetic correlations among these groups, encouraging the speculation that genetics and linguistics are linked.

Despite the uncertainties that surround attempts at deep reconstruction of language phylogeny, the tight correlation at the level of language phyla and genetically cohesive ethnic populations is an important discovery. The coincidence in pattern of genetic distances between major populations and archeological evidence for a recent African origin of modern humans joins that discovery and strengthens the model.

Recently Cavalli-Sforza and his colleagues collaborated with Kenneth and Judith Kidd of Yale University in collecting DNA samples from global populations with the same aim of producing an atlas of human variation. These data, which offer higher quality information than classic markers, so far cover just one-hundredth of the population size sampled by the original study; nevertheless, the DNA data agree closely with the protein data.

demanded by the multiregional model, he concludes. Cavalli-Sforza agrees. "Proponents of the multiregional model simply do not understand population genetics," he states. "They use a model that requires continuous exchange of genes, but it requires enormous amounts of time to reach equilibrium. There has been insufficient time in human history to reach that equilibrium."

Ecological Arguments

Evolutionary ecologists are no less critical of the multiregional model, particularly of its putative prime mover. "Recourse to culture as an all-embracing and all-pervading explanation has probably done most to obscure the processes by which modern humans evolved," argues Robert Foley, an ecologically oriented anthropologist at Cambridge University. "The origins of anatomically modern humans have been treated as a unique event, outside the context of the processes of evolutionary biology." The event will be best understood if it is examined in a "comparative ecological framework," he says.

The first point Foley makes is that anthropologists probably have been somewhat misled by their own nomenclature, specifically in the evolutionary shift from *Homo erectus* to archaic *sapiens* to modern *Homo sapiens*. The first of these two steps—from *erectus* to *sapiens*—involves going from one species to another, an apparently more significant change than going from archaic *sapiens* to modern *sapiens*, which seems merely to be modification within a species. In fact the change from archaic to modern forms of *Homo sapiens* is no minor anatomical shift, but involves a radical alteration in direction—a move away from the robusticity associated with thick cranial and postcranial bones; a reorganization of cranial and facial proportions with the raising and shortening of the cranial vault and the tucking under of the face. This change, plus the dramatic

new forms of behavior that followed the appearance of modern humans, implies an evolutionary event of greater significance than the *erectus* to *sapiens* transition—so radical, says Foley, that it cannot be viewed as "a simple case of continuously gradual change."

Foley suggests that anthropological nomenclature should reflect the biological significance of these changes. Currently, many anthropologists recognize several subspecies of *Homo sapiens*: *Homo sapiens sapiens*, for instance, is used to refer to modern humans, while archaic precursors include *Homo sapiens neanderthalensis*, *Homo sapiens heidelbergensis*, *Homo sapiens rhodesiensis*, and *Homo sapiens soloensis*. The precursor of all these subspecies of *Homo sapiens* is *Homo erectus*. To emphasize the relatively small evolutionary change between *Homo erectus* and the archaic *sapiens* subspecies—and the more biologically significant transition between archaic populations and modern humans—the *neanderthalensis*, *heidelbergensis*, *rhodesiensis*, and *soloensis* populations should be made subspecies of *Homo erectus*, while *Homo sapiens* would be reserved only for modern humans. Taxonomic nomenclature is always slow to change, and so far Foley's proposal has not been taken up. As already indicated, the most likely change to be adopted, if not formally, is to recognize the Neanderthals as a species distinct from *Homo sapiens*: *Homo neanderthalensis*.

Ian Tattersall, an anthropologist at the American Museum of Natural History who has also examined the biological status of modern humans and their forebears, suggests that the term "archaic *sapiens*" be dropped, calling it a "ragbag with no evolutionary meaning." The distinct anatomy of the varied populations should be recognized as designating different species: "something anthropologists wouldn't hesitate to do if they were dealing with any creature other than humans."

Morphological differences between closely related species of primate are commonly small and restricted to just a few characters. If anthropologists employed the same kind of criteria that are used for recognizing different species of primate they would be forced to admit three or four, perhaps more, distinct human species half a million to 100,000 years ago—*Homo neanderthalensis* being one. Not entirely facetiously, Tattersall attributes anthropologists' penchant for placing all archaic *sapiens* populations into the same evolutionary "ragbag" to a "generous liberal sentiment that leads to the inclusion in *Homo sapiens* of all hominids whose brain size falls comfortably within the modern range."

Evolutionary ecology can address the most likely mode of origin of modern humans, drawing on principles deduced from other animals in general and large primates in particular. Populations that are geographically widespread, as archaic *sapiens* was, tend to accumulate local genetic differences that sometimes lead to speciation. This occurs despite parallel evolution of certain functional features, such as aspects of locomotion or dentition related to diet. The evolutionary radiation of colobine and cercopithecine monkeys in Africa during the past 5 million years is a good example. If archaic humans followed an evolutionary pattern typical of large primates, speciation is the most likely mode of the origin of modern humans from one of their widespread populations. The next question is, which continent offered the most propitious ecological conditions for speciation?

Climatic change is widely held to be important in driving evolutionary change, specifically through reshaping local environments. Continuous forest cover that is fragmented through a cooler or drier climate may create isolated populations that, as explained above, may become genetically distinct from each other. Similarly, open country may become broken up by areas of forest in wetter climes, again with the potential of creating isolated populations and subsequent genetic differentiation. If such environmental changes occur too rapidly, however, extinction is a more likely result than speciation, when populations have insufficient time in which to adapt.

The period between 150,000 and 10,000 years ago experienced sharp environmental fluctuations as a result of the Late Pleistocene glaciation. The impact on African environments was considerable and is thought to have been responsible for the burst of speciation among monkeys of the genus *Cercopithecus*. Earlier environmental shifts are also considered to have been responsible for subspeciation events in chimpanzees and gorillas. But is there any reason to suppose that Africa rather than Asia or Europe might have produced more speciation opportunities during global climatic shifts? Yes, argues Foley, because the climatic changes that produced speciations in Africa produced much more rapid environmental fluctuations in temperate latitudes, making extinction more likely there. The sporadic occupation of Eurasia during this period may be indicative of the magnitude of its environmenal changes.

Consideration of the origin of modern humans in the context of principles of evolutionary ecology therefore offers support for a single origin model, with Africa as the setting. If the possession of "culture" (in effect, a somewhat limited range of stone-tool technology) altered the mode of evolutionary change, perhaps these principles do not apply. But this argument must be recognized as special pleading with no empirical basis.

Discrete Origin Favored

So far we have considered mainly pattern and process in modern human origins. To most observers, the weight of evidence—from fossils, genetics, and other aspects of biology—favors a discrete origin of modern *Homo sapiens*, in Africa or possibly the Middle East. As modern humans expanded into the rest of the Old World they effectively replaced many, perhaps all, existing archaic humans. In the next chapter we will explore some aspects of behavior, including behavioral differences between archaic and modern humans, which may in turn illuminate possible modes of replacement.

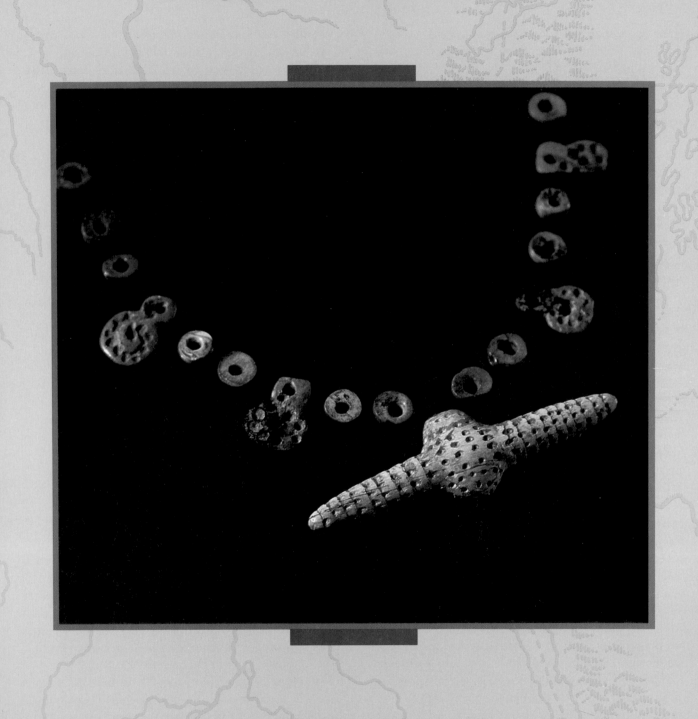

5

THE ARCHEOLOGY OF MODERN HUMANS

Evidence from physical anthropology and genetics ought, in principle, to generate conclusions about when and where modern humans evolved and what anatomical changes arose in the process. The role of archeologists is to reconstruct changes in behavior associated with this event. This task once seemed relatively easy, but as archeological evidence accumulated for regions outside Europe the pattern of change began to look much more complex and difficult to interpret.

In the European prehistoric record—specifically in southwest France and northern Spain—an abrupt transition occurred between 40,000 and 30,000 years ago. Stone-tool technologies, essentially unchanged for almost 200,000 years, were suddenly replaced by more sophisticated, stylish, and rapidly evolving artifact traditions. Body decoration (in the form of beads, pendants, and possibly necklaces) and artistic expression (in the form of engraved and painted

An Upper Paleolithic beaded necklace of mammoth ivory from Malta in Siberia.

images on objects and on cave walls) appeared for the first time. Long-distance trade and political connections arose. This new era, the Upper Paleolithic—clearly the product of the modern human mind at work—replaced the earlier era, the Middle Paleolithic. Until three decades ago scholars assumed this dramatic signal in the record to indicate that the modern human mind had arisen in Europe. A similar pattern of cultural change, perhaps delayed by a few millennia, was expected one day to be discerned in other parts of the Old World: a crisp, consistent pattern throughout.

As archeologists during the past three decades have folded back the pages of human history in other parts of Europe, in Asia, and in Africa, however, consistency of pattern has been distinctly lacking. A complex, confusing pattern is to be seen—incomplete, no doubt, but one that must have encrypted within it the story of the origin of the modern human mind. "Are we dealing with a process of gradual, cumulative change in the total range of cultural expression?" asks Cambridge University archeologist Paul Mellars. "Or with some kind of radical shift in the innate biological capacities of the human brain to accumulate and organize culture?" Perhaps because the still incomplete archeological record is equivocal at best, scholars respond to these questions in very different ways.

"The most economic explanation for the behavioral transformation [with the appearance of modern humans] is that it was grounded in the last of a long series of biologically based advances in human mental and cognitive capacity," argues Richard Klein, of the University of Chicago. "The special importance of this final advance was that it launched the fully modern human ability to manipulate culture as an adaptive mechanism." By contrast, Olga Soffer of the University of Illinois emphasizes social context as the driving force behind the advent of modern human cultural adaptation. "The differences between the Middle and Upper Paleolithic in northern Eurasia lay not in blade vs. teeth or in any morphological hardware, but in a dramatic change in economic and social relationships," suggests Soffer. "It is not a question of differences in *capacities* between archaic and modern people, but of *performance*." An analogy here, also supported by Ofer Bar-Yosef of Harvard University, is the shift in cultural context and subsistence that occurred with the agricultural revolution (and, of course, with the subsequent industrial and technological revolutions): qualitatively, the same people doing new and more complex things.

A biological (specifically, cognitive) change, or a shift in social organization? These are the two polar perspectives on the origin of modern human behavior. Different in implied mechanism, these two views also encompass different perspectives of our immediate ancestors. The first represents a real evolutionary event; the second lies in the realm of cultural revolution. If a real evolutionary event underlay the origin of modern human behavior, then our ancestors were archaic because they lacked the cognitive capacity to elaborate more complex behaviors. If cultural revolution initiated modern human behavior, then our immediate ancestors were archaic only because they had yet to launch that revolution.

Occupying a middle ground between these opposed perspectives is the notion that modern human behavior is a pluralistic, not a unitary, phenomenon, and that therefore its origins are likely to be complicated. "It would be foolish to assume the evolution in tandem of the set of biological and behavioral traits that anthropologists see as characteristic of modern humans," argue Philip Chase and Harold Dibble of the University of Pennsylvania. "It seems more likely that different biological and behavioral traits, even though they may be linked today, had functionally and temporally separate origins." Lawrence Straus of the University of New Mexico makes a similar point. "There was no *one* transition everywhere, even in as small an area as western Europe," he says. "There was no necessarily unified 'package' of Cro-Magnon anatomy, Upper Paleolithic stone tool technology and typology, bone/antler tools, large-scale specialized hunting, ornamentation, and art. Some (but not necessarily all) of these elements can be found in various combinations at different times and places; each region had its own history of changing circumstances and adaptive changes."

Patterns of Change

Patterns are always important in biology. In addition to the archeological pattern left behind by modern human behavior, there are the pattern of anatomy and the pattern of genetics. Concordance among the three patterns would encourage confidence that the fundamental events can be readily interpreted. As we saw in previous chapters, some scholars see reasonable agreement between genetic and anatomical evidence, specifically supporting the Out of Africa hypothesis. From anatomical evidence, modern humans appeared to have arisen first in Africa, some 100,000 years ago, and then expanded into the rest of the Old World over the next 60,000 years, replacing at least some existing archaic populations. Genetic evidence—from mitochondrial and nuclear DNA—can be interpreted to give the same overall pattern, even implying complete replacement of archaic populations. If modern human anatomy equates with modern human behavior, then evidence for that behavior should appear first in Africa and subsequently flow into the rest of the Old World. Concordance of patterns would be complete.

To examine this possibility, we will first build up the archeological pattern by looking at the Old World, region by region. Second, we will seek some insight into modern human social milieu and subsistence activities from archeological and anatomical evidence. Last, we will explore some of the evidence for interaction among modern and archaic populations, including ideas concerning mechanisms of replacement of one population by another.

A Sparse African Record

For reasons having to do with the history of the science, not only is the terminology of Old World archeology more complicated than it need be, but the field's overall perception is also dominated by southwest European evidence. The European record for this stage in human history is far more extensive than that for any other region—partly as a consequence of favorable preservation through archeological time, but in addition because Europe has been the target of far more archeological endeavors than has any other region of the world. This point cannot be overstressed. In terms of numbers, Africa has perhaps half a dozen good archeological sites from the Later Stone Age. Asia has perhaps 20 times that number, while Europe's record is at least 200 times richer than Africa's. The hegemony of the European archeological record is powerful and insidious. Some of the pattern seen in the European record may well apply elsewhere, but many of its details are likely to be idiosyncratic.

The shift between archaic human behavior and modern human behavior, labeled in Europe the Middle to Upper Paleolithic transition, is based specifically on a sharp change in the character of tool technology. The same terms are applied in Asia and North Africa; in sub-Saharan Africa, however, the equivalent stages are called the Middle Stone Age and Later Stone Age. So much for the jargon; but what do these terms actually describe?

Stasis Is Dominant

The Middle Paleolithic (Middle Stone Age) began about 250,000 years ago, ushered in by an important innovation in preparing the raw material from which artifacts were made. Known as the Levallois technique, after the suburb in Paris where the first specimens were recovered, it involved removing flakes

The Levallois technique, which first appeared some 250,000 years ago, allows the manufacture of many flakes of a standard shape and size, which can be further shaped.

Washington University. The technological transition coincides with the biological shift from *Homo erectus* to so-called archaic *sapiens*. "To use prepared core technology you not only have to understand stone very well, but you also have to have a very clear concept in your mind of what the finished shape is going to be," explains Brooks.

When referring to the range of implements manufactured by prehistoric societies, archeologists use several different terms, some of which have strong connotations. For instance the word "culture" (as in the expression Acheulian culture, for instance) is value-laden because it may be taken to imply a range of behaviors that go beyond the production and use of simple stone tools. Similarly, the word "industry" has references for the twentieth-century mind that may be inappropriate when dealing with simpler societies. Technology, as an item of vocabulary, is more neutral, giving a sense simply of the production and use of tools. Most neutral of all is the word "assemblage," which implies no specific cognitive capacity or other aspects of behavior in prehistoric people.

Although the Middle Paleolithic contained a larger range of tool types than the preceding Acheulian and included far finer implements, overall the new technology displayed one characteristic in common with its predecessor: stasis. There is some variability through time, of course, as there is over geographical region, but no major innovations. Some of these variants are given specific names; the most notable is the Mousterian technology, associated with the Neanderthal population of Europe and the Near East. Nevertheless, very few implements were made around 40,000 years ago (the end of the Middle Paleolithic) that were not already present near its inception. As the great French prehistorian François Bordes said of the Neanderthals: "They made beautiful tools stupidly." If the origin of the human mind is to be understood, it is important to identify signals of distinctly nonhuman behavior. Lack of innovation is surely one of them.

from a lump of rock to form a rounded shape with a flat top—the prepared core. Repeated striking around the periphery with a hammerstone could then yield many flakes of similar thinness, and the shape of the prepared core determined to a large extent the shape of the flakes produced from it. In addition to Levallois flakes, Middle Paleolithic industries often included side-scrapers, backed knives, hand axes, denticulates, and points. The shift from the long-established Acheulian industry to the Levallois-based Middle Paleolithic was "a cognitive transition," suggests Alison Brooks, an archeologist at George

The Concordance of Patterns Fragments

Then came the Upper Paleolithic. In Europe the rich prehistoric record documents an unprecedented variability in artifact assemblages, both geographically and temporally. In the 1950s, Bordes argued that this variability represented stylistic intent in different ethnic groups. A decade later American archeologists Sally and Lewis Binford suggested that the variation was the result of functional, not ethnic, differences. For instance, sites occupied for a matter of days or even weeks would involve a range of maintenance functions different from the functions carried out at kill or butchery sites; these functions would be reflected in different tools made and used. The functional differentiation hypothesis has not stood up, however, and these days archeologists prefer something more akin to Bordes's original proposal. "My own suggestion is that these changes in artifact forms should be seen as a reflection either of personal 'style' in artifact manufacture or of the emergence of more complex and highly structured cognitive systems associated with the appearance of fully developed language," suggests Paul Mellars. As we shall see, the notion that fully modern language is at least one—if not the sole—cognitive innovation driving the origin of modern humans is popular and persuasive.

In any case, the variability of artifact assemblages through time and space is now regarded as indicating a real change in level of innovation at the beginning of the Upper Paleolithic. An explosion in the creativity of tool production and an expansion of raw materials and new traditions followed closely on each other. Innovation was now measured in a few thousand years, not the hundreds of millennia of earlier eras.

The European Upper Paleolithic began with the Aurignacian tradition, some 40,000 years ago, and included what amounts to the signature of early modern human stone technology: a proliferation of nar-

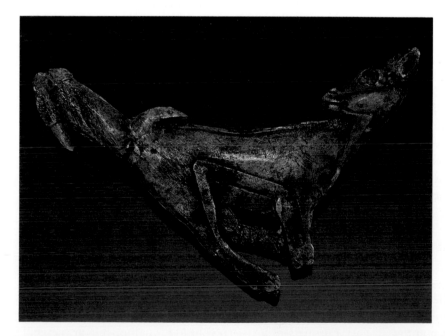

Antler spear-thrower, showing a young ibex with emerging terd, on which two birds are perched; from the cave of Mas d'Azil, in the Ariège region of France, dated to the Magdalenian period.

Head of a spear-thrower, showing a bison licking its flank; from the cave of La Madeleine in the Dordogne region of France, dated from the Middle Magdalenian period.

row blades. Upper Paleolithic people made many fine tools from bone, antler, and ivory, something that was uncommon in earlier times. An important aspect of Upper Paleolithic tools is that classification into types, based on consistency of form, is much easier than with Middle Paleolithic tools; Upper Paleolithic people appear to have had a clear notion of the intended final product, and the ability to fulfill it. In addition, the Aurignacians produced large quantities of beads, presumably for personal adornment. If it is true that modern humans came into Europe from elsewhere, then the Aurignacians represented that first wave, and it was they who must have encountered established Neanderthal populations. Later we will see that there is evidence for contact between these two different peoples.

The Aurignacian was followed in succession (with some geographical differences) by the Gravettian, Solutrean, Magdalenian, and finally the Azilian peoples, each producing stylistic variants upon the Upper Paleolithic theme. Musical instruments began to be made, and painted and engraved images and objects became a steadily more important element of some kind of social context, presumably ritual in some still elusive sense. Accomplished hunters, Upper Paleolithic people lived in larger settlements than occurred in the Middle Paleolithic, perhaps reflecting more sophisticated economic and social systems. Evidence of long-distance transport of both utilitarian (stone) and nonutilitarian (shells and amber) objects indicates the establishment of large-scale alliances like those of modern foraging peoples. "In most small-scale societies, exchange, or trade, operates as a vehicle of social obligation," observes Randall White, an archeologist at New York University. "Obligations are social bonds capable of tying together different social groups."

For White, as for many (but by no means all) anthropologists, the appearance of modern humans in Europe was a true revolution, qualitatively and quan-

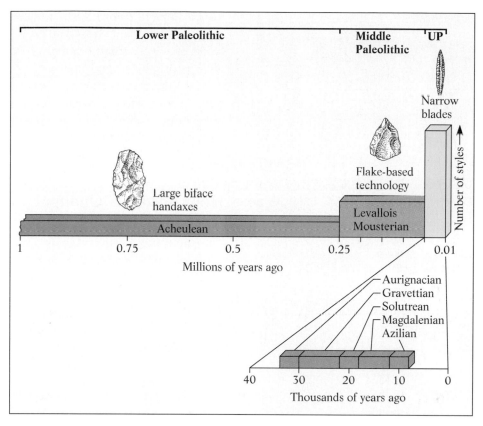

Stasis marks much of the history of stone-tool manufacture, with the number of discernable implements beginning to rise appreciably with the arrival of archaic sapiens *and particularly so when modern humans appear. Between 40,000 and 10,000 years ago (the Upper Paleolithic) styles change rapidly, indicating the new influence of innovation and cultural norms.*

titatively. "All [the evidence] is consistent with the idea of a total restructuring of social relations across the Middle/Upper Paleolithic boundary," he says, "in the course of which corporate and individual identity became important and are enhanced by stylistic input and regional differences in working of stone, antler, and bone, the fabrication and wearing of ornaments, and the regular aggregation of a set of otherwise dispersed local groups." Richard Klein puts it this way: "In the broad sweep of European prehistory, [the Upper Paleolithic people] were the first people for whom archeology clearly implies the presence of both

'culture' and 'cultures' (or ethnicity) in the classic anthropological sense." If this kind of characterization is accepted, then in Europe the anatomical and archeological patterns match: modern human behavior accompanies the arrival of anatomically modern humans.

With Asia the neat concordance of patterns begins to fragment. In the East, where the prehistoric record is extremely sparse, there is no clear Middle to Upper Paleolithic transition, in the sense of a switch from flake to blade technologies. A crude flake-and-chopper technology, accompanied perhaps by the use

Richard Klein of the University of Chicago has worked extensively with excavations in southern Africa, looking for evidence of shifts in subsistence with the appearance of modern humans.

the incomplete record. "The impression of remarkable continuity and conservatism in the Far East is based on a very small number of excavated sites, often poorly dated and unevenly described," he cautions. "Wherever the archeological evidence is reasonably complete and well-dated [in the Old World in general], it implies that a radical transformation occurred about 40,000 years ago."

How Abrupt a Change?

The pattern in western Asia is different, more puzzling, and in many ways more interesting. In the Middle East, for instance, Neanderthals and anatomically modern humans appear to have coexisted from about 100,000 years ago to perhaps 50,000 years ago. Such a pattern effectively precludes an ancestor–descendant relationship between them; neither is there incontrovertible anatomical evidence of genetic interaction (interbreeding) between these peoples. The puzzle, however, is that from all the archeological evidence so far recovered—and it is sufficiently extensive to encourage confidence in the observed pattern—there was no technological difference between the Neanderthals and their anatomically modern human neighbors, at least in the range of artifacts they manufactured. Both peoples apparently produced rather typical Mousterian-like industries, a link not surprising for the Neanderthals but certainly challenging for expectations regarding anatomically modern humans.

"The strong impression is that the local populations of archaic and modern humans were able to coexist in some kind of reasonably balanced equilibrium in which the anatomically more 'advanced' forms had no very obvious competitive advantage (in a demographic sense) over the anatomically modern ones," observes Paul Mellars. He wonders whether the modern humans, having evolved under equatorial conditions in Africa, would indeed have been ill-equipped "for large-

of bamboo "blades," continued until about 10,000 years ago (the beginning of the Neolithic period), according to Geoffrey Pope of the University of Illinois. Pope and others also argue that the appearance of bone artifacts, art, and sophisticated burials— signals of the modern human mind at work—was a gradual, not abrupt, process in East Asia. However, Klein wonders whether this picture is an artifact of

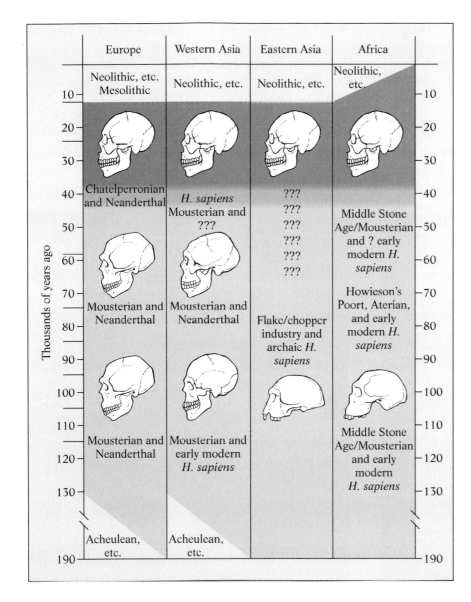

	Europe	Western Asia	Eastern Asia	Africa
10	Neolithic, etc. Mesolithic	Neolithic, etc.	Neolithic, etc.	Neolithic, etc.
20				
30				
40	Chatelperronian and Neanderthal	*H. sapiens* Mousterian and ???	??? ??? ???	Middle Stone Age/Mousterian and ? early modern *H. sapiens*
50			???	
60			??? ???	
70	Mousterian and Neanderthal	Mousterian and Neanderthal		Howieson's Poort, Aterian, and early modern *H. sapiens*
80			Flake/chopper industry and archaic *H. sapiens*	
90				
100				
110	Mousterian and Neanderthal	Mousterian and early modern *H. sapiens*		Middle Stone Age/Mousterian and early modern *H. sapiens*
120				
130				
190	Acheulean, etc.	Acheulean, etc.		

Thousands of years ago

The paleontological and archeological records of the origin of modern humans differ widely between Africa, Europe, and Asia. Some of this reflects real variability in prehistory, while differences in availability of evidence is also a factor.

scale colonization of the very different environments of the northern latitudes." Klein also points to the possible equatorial origin of the anatomically modern humans of Skhul and Qafzeh to explain the fact that they had "no obvious behavioral (cultural) advantage over Neander-

thals." The reason, he says, could be that Neanderthal bodies were stockily built, with relatively short limbs (an adaptation to cold), whereas early anatomically modern humans were tall, with relatively long limbs (an equatorial adaptation).

However, perhaps anthropologists have been misled by their own terminology and by assumptions flowing from historical description. The terms "modern humans" and "anatomically modern humans" are often used interchangeably in discussions about the origin of the kind of people who occupy Earth today. Yet the only sure signs of the presence of modern humans are the products of the modern human mind noted in this chapter. The search in the fossil record for anatomically modern humans—judged by the shape and robusticity of the skull and the rest of the skeleton—may be only part of the story of modern human origins. "Both cranially and postcranially, [the people of Skhul and Qafzeh] are clearly far better ancestors for later modern people than the Neanderthals were," argues Klein, "but it seems reasonable to

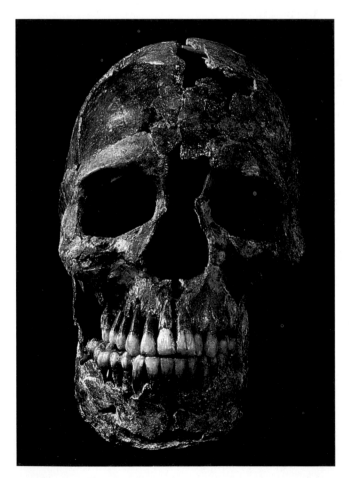 

Neanderthals and anatomically modern humans apparently coexisted in the Middle East for as much as 60,000 years; here we see the skull Qafzeh 9 (left) which is anatomically modern and may be as much as 100,000 years old, while the Neanderthal from Amud (right) is some 20,000 years younger.

suppose that the reason their behavior was not fully modern is that they were not yet fully modern biologically, perhaps above all neurologically." In other words, the modernity of these early anatomically modern humans may only have been skin deep, almost literally. Their skeletons had begun to take on the more gracile characteristics of later modern people, but they may have lacked a key evolutionary change in cognition, presumably one that underlies language.

This apparent uncoupling of early modern human anatomy from modern human behavior is unsettling because, although it is sometimes difficult to interpret, external morphology of modern people is tangibly distinguishable from that of archaic people. Evidence of the modern human mind at work is often much less so. When confronted with, say, the tool technologies of the Upper Paleolithic or the painted and engraved images to be found in both Europe and Africa, no one will question that they are the work of modern humans. But the nature of complex human behavior is such that the absence of such evidence is not necessarily evidence of absence of the modern mind. "We are going to have to be much cleverer in understanding what the archeological record can tell us," says Alison Brooks in response to this challenge.

The challenge is particularly demanding in Africa, partly because the pattern in some ways is like that in the Middle East but also because the archeological record itself is extremely sparse, far more so than most people suppose. As in the Middle East, anatomically modern human fossils have been identified at an early date (close to 100,000 years ago) in sub-Saharan Africa. Again echoing the Middle East, the archeological record of sub-Saharan Africa does not bespeak a dramatic shift from archaic to modern behavior coincident with the appearance of anatomically modern humans. That shift, from Middle to Later Stone Age (like the Middle to Upper Paleolithic transition), may have occurred close to 40,000 years ago. Qualitatively, the sub-Saharan African event was

very similar to what happened in Europe (and in north Africa): a strong emphasis on blade technology, the inclusion of bone and ivory as raw materials for artifacts, and the production of body ornamentation of painted and engraved images.

The modern human behavioral package was definitely in place in Africa 40,000 years ago, even though it varied somewhat from the version in Eurasia. Prior to that date—at least according to some observers— evidence of modern human behavior in sub-Saharan Africa that eventually would be exported to Eurasia is uncertain. Two industries of particular interest in this respect are the Aterian (in northwest Africa) and Howieson's Poort (in southern Africa). Both are essentially Middle Stone Age in character, but include unusual elements: in the Aterian, for instance, some points and scrapers have small "stems" on them; in Howieson's Poort, small crescent artifacts are "backed" (blunted along one edge). There is some speculation that these modifications may have been for making hafted tools: a stone segment secured into a wooden holder. If this interpretation is correct, these would be the oldest composite tools known.

The more general question here is over the significance of the two industries. Do they represent nothing more than geographic variants on Middle Paleolithic and Middle Stone Age technology? Or are they the product of a new kind of mind at work, the modern human mind that eventually yielded the fully expressed Later Stone Age and Upper Paleolithic levels of innovation? Klein, for instance, suggests that claims for an archeological signal of modern human behavior in these artifacts inflates their significance. Apart from these peculiarities, "neither industry is meaningfully distinguished within the more general Middle Stone Age/Middle Paleolithic complex," he says. He also points out that the Aterian and Howieson's Poort industries are succeeded by typical Middle Stone Age, Middle Paleolithic assemblages, in which case they cannot be seen as transitional to fully modern behavior.

*I*nvestigating *Subsistence Skills*

Did modern humans exploit the resources of their environment more efficiently than their hominid predecessors? Testing such an assumption presents several challenges. The most systematic and successful attempt to address the issue has been the comparison of Middle Stone Age and Later Stone Age sites in southern Africa by Richard Klein. His conclusion is that the subsistence skills of modern humans were demonstrably greater than earlier peoples'.

Fundamental differences in subsistence capabilities may be reflected in the nature of the debris left behind at living sites, and the anthropologist's task is to interpret their significance. There are, however, many reasons other than differences in human activity that might impose material differences on living-site debris. The most common confounding factor is the fact that carnivores, particularly hyenas, may make their dens in the same sites that humans occupy at other times; it may then prove difficult or impossible to determine which bones were brought to the site by humans and which by hyenas.

Even when there is no potential obfuscation by carnivore activity, differences in fossil assemblages at living sites may still be the result of nonhuman activity. For instance, fluctuations in the relative abundance of reindeer versus red deer at living sites in France from about 700,000 years ago to 250,000 years ago appear to be the result of climatic shifts: reindeer were abundant during cold periods, red deer during warmer periods. Klein has shown that during cool, dry periods in southern Africa grazing animals were more common on the land and in living sites, while browsers dominated during warmer, wet times.

For a reliable comparison between living sites separated by many thousands of years, it is essential that similar environmental conditions prevailed. Klein identified two such sites in the southern Cape Province of South Africa, Klasies River Mouth Cave and Nelson Bay Cave. The former has Middle Stone Age (MSA) living sites 130,000 to 115,000 years old, while the latter has Later Stone Age (LSA) remains close to 10,000 years old. Both represent interglacial times, with essentially the same fauna, including coastal resources. Klein spent several years excavating the sites in the early 1970s and even longer analyzing the data.

There are two striking contrasts between the sites. First, the bones of marine fish and seabirds (cormorants, gulls, and gannets) are abundant at the Later Stone Age site but rare at the Middle Stone Age site. Klein concludes that Middle Stone Age people had not mastered fishing and fowling techniques. Grooved stones that may have been for net or line sinkers are present among the Nelson Bay (LSA) remains but absent from the Klasies River Mouth (MSA) assemblage. Similarly, toothpick-sized, double-pointed bone splinters that could have been used as hooks for catching fish or marine bones are present at Nelson Bay but not at Klasies River Mouth.

The second difference is in the proportion of potential prey species. The Nelson Bay fossil remains roughly match the historical abundance of species, while those at Klasies River Mouth are heavily biased against dangerous animals and toward pliant animals. For instance, the Middle Stone Age people at Klasies River Mouth Cave depended heavily on eland, which are known for their docility and the ease with which they can be herded. Cape buffalo and bushpig, by contrast, are notoriously dangerous, particularly when cornered: their bones are uncommon at Klasies River Mouth. Klein suspects that Later Stone Age people had developed techniques for killing at a distance—with bows and arrows, for instance, and snares. Archeological evidence of bows and arrows, in the form of small stone points that could be set in shafts, is known from the region from sites 20,000 years old.

Klein was concerned not just with the numbers of individual species at each site, but also with their sex and age. Differences in shape or size of horns would be the best indicator of sex for bovids in the assemblages. Unfortunately, for various reasons horns are uncommon. The large size of males compared with females in many species also offers a potential means for determining the sex of the

The Klasies River Mouth Cave in southern Africa was occupied during Middle Stone Age times, when hunters were apparently less proficient than later, modern humans.

remains. Measurements on large numbers of the same skeletal element for a species should theoretically yield a bimodal distribution in size, with males in the large-sized group and females in the small-boned group. The fragmentary nature of the fossil remains frustrates this line of approach, too, as many bovid bones are rather similar among different species and difficult to differentiate, even when whole.

Determination of age proved more tractable. Measurement of the crown height of molar teeth yields information about an individual's age, as teeth gradually wear down throughout an animal's life. By this means

Klein discerned two patterns in the fossil remains. For eland and bastard hartebeest, another docile animal, the age profile of the individuals at Klasies and Nelson Bay reflects the age profile of a standing herd, a pattern known by paleobiologists as catastrophic (the implication is that a herd or part of a herd met its end catastrophically, perhaps by being driven over a cliff edge). The second pattern, known as attritional, shows a great preponderance of very young and old animals, with few prime-age individuals. The attritional pattern reflects the different vulnerabilities of individuals of different ages, and it matches the profile that effective car-

nivores such as lions are able to take from dangerous prey species, such as buffalo. In the Klasies River Mouth and Nelson Bay assemblages the species with attritional profiles are Cape buffalo, blue antelope, roan antelope, and giant buffalo—all potentially difficult prey.

While Klein infers from the fossil evidence that Middle Stone Age people were hunters, albeit less effective than later people, Lewis Binford asserts that they were mostly scavengers, particularly of larger species. Binford bases his claim on the relative abundance of skull and foot bones and corresponding rarity of limb bones among the Klasies River Mouth remains. These body parts, low in nutritional value, would be what remained available to a poorly armed scavenger, he argues; the meat-bearing limb bones would be taken by the primary scavenger. Klein points out that this pattern is not confined to Middle Stone Age sites but is found much later, even on African Iron Age sites where the animals in question were cattle—most definitely not scavenged. Skull and foot bones are dense compared with other parts of the skeleton, and therefore more likely to be preserved. Despite the challenges of interpreting living-site debris, Klein further points out that the human fossils associated with the Middle Stone Age assemblage at Klasies River Mouth Cave have been diagnosed as anatomically modern—in which case, he says, "the evolution of modern physical form preceded the evolution of fully modern behavior."

Alison Brooks disagrees, arguing that the Aterian and Howieson's Poort assemblages must be seen as distinct from Middle Stone Age, Middle Paleolithic technology. This belief is based in part on the occurrence of ostrich shell worked as beads at several sites in the 40,000–90,000-year range. In addition, Brooks and fellow archeologist John Yellen have discovered finely crafted harpoon points made from bone at two sites in Zaire. Very similar in appearance to 14,000-year-old objects in European Upper Paleolithic sites, those from Zaire may be as much as 90,000 years old. If so—and there is considerable uncertainty about the date—the presence of

If this harpoon point from a site in Zaire is as old as some archeologists suspect—as much as 90,000 years—it would indicate modern human behavior at a very early time in Africa.

the harpoons would indicate a much earlier appearance of modern human behavior in Africa than in Europe. Overall, says Brooks, there is evidence for "the beginning of a change in behavior about 80,000 or 90,000 years ago in sub-Saharan Africa."

What Brooks does not see, however, is an obvious link between what was developing in Africa and what later appeared in Europe in the Aurignacian. "There is no archeological evidence of a wave of people flowing out of Africa and into Europe," suggests Brooks. "You get no support for population replacement from the archeological evidence." Klein, however, believes that the character of the Middle to Upper Paleolithic (Middle to Later Stone Age) transitions provides reasonable support for some replacement. "The relatively abrupt and radical nature of the change provides far better support for the 'Out of Africa' hypothesis than for its multiregional alternative," he says. "I think the basis for replacement was the competitive edge conferred by the fully modern human ability to wield culture. The question remains where did this competitive edge arise. Based on present archeological and fossil evidence, only Europe can be ruled out." Of the remaining contending regions— the Middle East and Africa—Africa seems more likely, "but the archeological record provides only partial support."

Not surprisingly Alison Brooks, among others, challenges this perspective. "A closer scrutiny of the archeological record leads one to inquire, Just how abrupt *was* the behavioral transition in Europe?" she says. "I believe that the gulf between the Middle Paleolithic and the Upper Paleolithic has been artificially widened by deemphasizing the very real evidence of cultural complexity in the former and overstressing the achievement of early modern humans, who, in Europe, did not achieve all of the behaviors usually cited as part of the Upper Paleolithic 'revolution' until the very end of the Pleistocene [near 10,000 years ago]." Geoffrey Clark and John Lindly of Ari-

Alison Brooks and John Yellen, archeologists at George Washington University and the National Science Foundation respectively, have made important finds at sites in Zaire, which may indicate the presence of modern human behavior much earlier in Africa than in Europe.

zona State University support Brooks's interpretation and equate the opposite view with "the recent rash of 'Neanderthal bashing' in the press." A more dispassionate reading of the archeological record, they contend, shows "clear evidence of cultural and behavioral continuity between the Middle and Upper Paleolithic." In addition, they say, the archeological record "shows none of the indications of abrupt discontinuity that would be expected had population replacement without admixture actually occurred."

Scholarly opinion, then, is as divided over the interpretation of the archeological evidence as it is over that of the fossils. Gradual behavioral change, or abrupt? Continuity or replacement? Familiar themes, ones that have been central to the debate over the origins of modern human populations throughout the greater part of this century. That the issues remain unresolved reflects the uncertainty involved in identifying unequivocally what the presence of modern human behavior looks like in the prehistoric record.

Scavengers or Hunters?

We now turn to aspects of the economic life of early modern humans, again making comparisons with earlier peoples. Our object is to discover "what were the selective advantages of the early modern human biocultural system that allowed it to become the sole pattern of human behavior after a relatively short period of evolutionary time," as Erik Trinkaus frames it. "Since the transition was one in which both human biology and culture were involved, the issue must be addressed in terms of human biological changes, cultural alterations, and the interactions between them." The focus in this chapter will be on biological factors.

As we have noted, the most striking anatomical difference between archaic and modern humans is that of robusticity, best documented among the Neanderthals, but generally true of all archaic humans. As bone morphology is, in large degree, a response to patterns of use, "this robusticity [of archaic humans] is a reflection of both strength and endurance," explains Trinkaus. The leg bone anatomy of Neanderthals implies that "they spent a significant portion of their waking hours moving continuously across the landscape." The more gracile build and relatively longer legs of early modern humans gave them "a significantly more efficient locomotor pattern" that involved a purposeful striding gait, says Trinkaus. Recently, Yoel Rak of Tel Aviv University has suggested that a key difference in pelvic anatomy between archaic and modern humans indicates that the latter were also more efficient runners.

In addition to their apparent changes in locomotion, early modern humans also used their hands differently, as evidenced by the reduced robusticity of the entire upper limb. There was a shift in emphasis from power to precision, a clear sign of increased

A comparison of fingertip bones from Neanderthals (right) and modern humans reveals how much more anatomically robust Neanderthals were; their facility for fine manipulation would have been limited.

manipulative skills, says Trinkaus: "The origin of modern humans saw a marked decrease in the habitual level of strength exerted by the human upper limb, a shift toward more precision use of the hand, limitation of heavy manual tasks to the power grip, more extended positions of the elbow and fingers, and changes in habitual grip positions." In other words, suggests Trinkaus, archaic humans' adaptive strategy was anchored in strength, while modern humans relied more on skill; clever use of tools took the place of brute force.

No one doubts that early modern humans were accomplished hunters, pursuing the ecologically powerful mixed economy of hunting meat and gathering plant foods. The evidence is abundant in the technology and the remains of animals eaten. There is less agreement, however, on how different the modern human subsistence strategy was from that of archaic humans. For instance, Lewis Binford of Southern Methodist University argues that modern humans virtually invented hunting as a hominid activity. "Between 100,000 and 35,000 years ago the faint glimmerings of a hunting way of life appear," he says. "Our species had arrived—not as a result of gradual, progressive processes but explosively in a relatively short period of time." Archaic humans were little more than opportunistic scavengers, grabbing what they could from carnivore kills, perhaps occasionally bringing down small prey, but doing nothing that could be described as systematic hunting. Archaic humans lacked the ability to plan ahead, argues Binford, a key skill in the complex social and economic organization that represents the hunter-gatherer way of life. Trinkaus agrees with this assessment, saying that archaic humans lacked "the planned subsistence patterns characteristic of recent hunter-gatherers and hence a high degree of minimally directed movement during the food quest." In other words, implies Trinkaus, archaic humans simply wandered around the landscape, taking whatever prey or carcasses they happened to stumble over. Binford's assessment is based largely on his analysis of animal remains at archeological sites, whereas much of Trinkaus's view derives from inferences from the robust nature of the archaic human skeleton.

"Binford's position is extreme on this issue," counters Richard Klein, who has made extensive analysis of faunal remains on archeological sites. "There's no doubt that archaic humans were less efficient hunters than modern humans. I can see this clearly in the African record." Klein's analyses show that modern humans had greater mastery of their environment, being able to take more dangerous and difficult-to-find prey and other food items, as we saw earlier. Unfortunately, no such systematic study has been completed on the faunal remains at sites in Europe and Asia. There are suggestions that animal bones from Middle Paleolithic sites are principally those of very young or old individuals and that bone assemblages that more closely reflect the age profile of standing herds occur only in Upper Paleolithic sites: the former is the signature of a scavenger, the latter that of a proficient hunter. In addition, some Upper Paleolithic sites are heavily biased toward one species, such as reindeer or ibex. This has been taken to imply a degree of systematic exploitation of resources that was absent earlier. "These conclusions may be correct," observes Klein, "but until the assemblages are studied more thoroughly we cannot be certain."

The debate, therefore, is over the magnitude of the behavioral change that occurred at the Middle to Upper Paleolithic transition. Was it incremental or radical? The first appearance of artistic expression and real technological innovation—as well as substantially more intense social milieu, manifested in larger living sites, a greater density of sites, and long-distance contacts—clearly speaks of significant change, which most anthropologists view as being biologically driven rather than cultural revolution. But the suggestion that archaic humans wandered aimlessly, utilizing what resources they happened upon, is not convincing. Chimpanzees are known to exploit their resources with a degree of purpose—knowing when a certain tree is likely to come into fruit, or

where to find water in particularly dry times. Sometimes a troop may divide in two in the morning, each foraging independently during the day but both reaching the same location in the evening, as if an agreement had been made.

With brains three times the size of those of chimpanzees, archaic humans are likely to have done at least as well, and probably substantially better, in planning their foraging activities. "If you look at the things archaic humans made with their hands, Levallois cores and so on, that's not a bumbling kind of thing," observes Margaret Conkey of the University of California at Berkeley. "They had an appreciation of the material they were working with, an understanding of their world. I'm not saying that they were just like us, but I think the differences have been exaggerated." The insufficiency of the archeological and paleontological evidence allows different interpretations. What is incontestable, however, is that archaic *sapiens'* subsistence activities required great strength and endurance.

A clear difference certainly existed, in the realms of culture and subsistence, between archaic and modern humans—at least when compared across the Middle to Upper Paleolithic boundary. Given such a difference, it is not difficult to imagine that modern humans might have enjoyed a considerable competitive edge through superior social and economic organization and technology. Assuming that the Out of Africa hypothesis is valid, how might this competitive edge be expressed as modern populations encountered existing local archaic populations? How, in the most extreme case, would incoming modern humans replace the local archaic people?

Evidence of Coexistence?

Supporters of the multiregional hypothesis frequently claim that population replacement must have been a bloody affair, an early example of genocide. Recall

Milford Wolpoff's characterization: "Rambo killer Africans, sweeping through Europe and Asia." It is impossible to imagine one human population replacing another, except through violence, he asserted. It is worth pointing out, however, that examples in recent history—the near-genocide of Native Americans and Australian Aborigines—were in the tradition of colonial occupations, with a long-established history of warfare behind the colonizers. Is it actually impossible to think of population replacement except by violent means? Not necessarily.

First of all, direct archeological evidence for contact between archaic and modern populations is nil. No unequivocal co-occurrence of archaic and modern remains, for instance has been found. As we shall see, indirect evidence exists that Neanderthals learned some new technology from Aurignacian people. There is, however, no archeological sign of violence.

The archeology of warfare fades fast as one looks back through human history, rapidly disappearing beyond the Neolithic (10,000 years ago), when agriculture and permanent settlements began to develop. In the earliest civilizations monumental architecture seems often to be almost a celebration of victorious battles with the enemy. Even in earlier times, between 5000 and 10,000 years ago, indications of violence can sometimes be seen in paintings and engravings. But go back beyond the beginning of the agricultural revolution and depictions of war in art virtually vanish. Again, the absence of evidence cannot be taken as evidence of absence, and the archeological record itself—given its degree of incompleteness—cannot be relied upon to furnish direct evidence of bloody clashes. Warfare, so important a part of human history, is associated with the need for territorial possession that arose once populations became agricultural and necessarily sedentary. Violence became an obsession once populations started to expand and the ability to organize large military forces developed. Is violence, then, a necessary and inevitable characteristic of humankind or merely an adaptation to certain circumstances? It may have occurred

when modern humans replaced archaic populations, but it is not yet possible to be sure one way or the other. What are the alternatives?

"There are many ways in which invading populations might wipe out local people," argues Chris Stringer. "Competition for resources, for instance, and the introduction of exotic diseases. There are plenty of historical examples of local populations collapsing through introduced diseases against which they have no immune defenses." One good example, of course, is the devastation that Pizarro's modest band of Spanish soldiers wreaked on the mammoth Inca Empire. Superior technology, including horses, was no doubt a factor; but the real killing agent was smallpox, a lethal viral disease that swept through the immunologically naïve population.

Historical examples of simple competition for resources are more difficult to come by, but Ezra Zubrow, an anthropologist at the State University of New York at Buffalo, has carried out some theoretical analyses, with surprising results. In the European context, he says, "the Neanderthals could have become extinct in a single millennium"—the pattern seen in the prehistoric record. A modest difference in subsistence skills—amounting to some 2 percent difference in mortality—hardly seems like the potential agent of population destruction. But very often in biology our perceptions are rooted in present experience, and we fail to appreciate the impact of small differences over relatively long periods of time. In this case, a slim margin in vitality over a millennium translates into one population succeeding while the other succumbs. Zubrow's results do not confirm that modern humans indeed replaced archaic humans simply through the impact of a slim advantage in resource competition. But his work establishes this as a plausible alternative.

Evidence that archaic and modern humans coexisted comes from two parts of the world, the Middle East and western Europe. As we saw earlier, archaics and moderns might have been contemporaries for as long as 50,000 years in the Middle East. We also saw that, in tool technology at any rate, there was no discernible difference in behavior between the two populations. This could mean that, nevertheless, there were "sufficiently different patterns of behavior and/or land use to maintain the two groups largely genetically separate for at least a thousand generations," as Erik Trinkaus puts it. Or it could be that the coexistence is more apparent than real. Ofer Bar-Yosef, for instance, considers it possible that archaic and modern populations in the Middle East never actually encountered each other. "From fossil evidence of microfauna you could imagine that as climate fluctuated you had movement of the two populations, in concert," he explains. "In colder times, populations would move southward, and the Neanderthals would occupy the region. When it warmed up, northward population movement would take Neanderthals out of

the region and modern humans would enter from the south"—a kind of climatic musical chairs.

While most anthropologists see no evidence in the fossil or archeological records of the Middle East that archaic and modern populations interacted in any way, in Europe, the story is different. At the Middle-Upper Paleolithic boundary in western Europe, a stone-tool tradition exists that is neither archaic nor modern, but a mix of both. Called the Chatelperronian, the industry includes some of the earliest bone, antler, and ivory implements. Because of its intermediate character, the Chatelperronian has been said to represent the work of a population that was evolving from Neanderthals to modern humans: an evolutionary intermediate. It has therefore formed part of the multiregionalists' argument. In 1979, however, François Lévêque of the University of Bor-

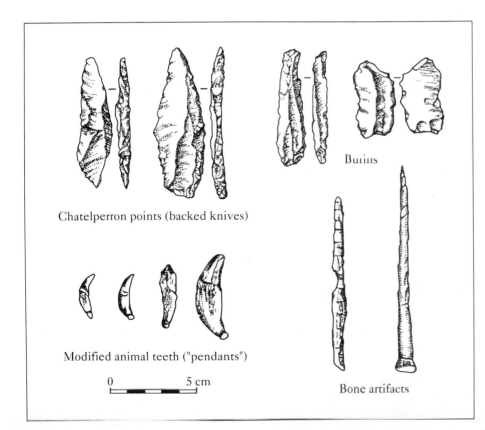

Chatelperron points (backed knives)

Burins

Modified animal teeth ("pendants")

0   5 cm

Bone artifacts

The Chatelperronian assemblage, which includes some of the earliest bone, antler, and ivory implements, is intermediate in character between Middle Paleolithic and Upper Paleolithic industries. Some archeologists suggest the industry was the work of a population that was evolving from Neanderthals to modern humans, while others argue that it was the result of contact between the two separate cultures and not an indication of evolutionary continuity.

*P*opulation *Extinction*

In Europe Neanderthal and modern human populations overlapped geographically and temporally, particularly in the west, for approximately two millennia: in central Europe Neanderthals disappeared perhaps as late as 38,000 years ago, and in western Europe about 32,000 years ago. Although the proximate cause of the extinction is unknown, its suddenness invites speculation that the process must have been dramatic.

In an attempt to circumscribe the possibilities, Ezra Zubrow has developed models that explore the outcome of an interaction of two populations under various demographic conditions. The principal conclusions are two: first, the population dynamics of interaction and outcome may be complicated, and sometimes counterintuitive; second, very small differences in the success of one population over another can produce unexpectedly large effects in remark-ably short periods of time. A difference of 1 percent in the subsistence abilities of *Homo sapiens* over *Homo neanderthalensis* in the context of competition over resources could lead to local extinction of the latter within 30 generations—less than a millennium.

Zubrow produced "models of interactive growth." The population dynamics depend on the fact that individuals of the two populations occasionally encounter each other and as a result affect each other. The potential outcomes of such encounters are several: 1) immediate withdrawal; 2) competition for resources; 3) warfare; 4) trade and exchange. Interactive models encompass all but the first of these possibilities.

Demographers describe populations in many ways, but an important tool is the life table, which effectively records the mortality rates for different age and sex cohorts in the population. Life tables may be used in two basic ways. The first is to represent the mortality experience of all cohorts of an entire population in a short period of time, such as a year: a snapshot of death for the population, as Zubrow phrases it. The second way is to follow the mortality record of a single age-sex cohort from birth to death. In this case one sees a trail of death, not a snapshot. For his analysis, Zubrow combined elements of both and plotted the fates of successive population cohorts through many generations.

A concept with which demographers have worked since the early years of this century is that of the stable population, introduced by population biologist Alfred Lotka. Stable populations may be of four different forms, defined by fertility and mortality profiles, termed West, North, East, and South—based in part on the actual demographic attributes of such populations in various parts of the world. For instance, the lowest fertility and mortality rates are represented in the Western group and the highest in the Southern group. Within each group there are 20 subtables, with graded levels of fertility and mortality.

deaux discovered a skeleton at St. Césaire, in southwest France: a Chatelperronian site. If the Chatelperronian industry had indeed been produced by a population in evolutionary transition from Neanderthal to modern, this was expected to show in the anatomy of Chatelperronian people.

"This was absolutely not the case," says Bernard Vandermeersch, a colleague of Lévêque. "The skeleton St. Césaire was most definitely a Neanderthal, very characteristically so." Later, remains of a second individual were found, also a Neanderthal. "You can imagine that the Neanderthals and the incoming modern populations had some contact, and that the Neanderthals learned some of the new technology," suggests Vandermeersch. Recent dating of this site and of others indicates that Neanderthals and

Left unperturbed, any one of the populations represented by these subtables would persist through the generations, their demographic characteristics unchanged. If you introduce a small increase in mortality in the first generation, however, decline may be triggered, with mortality increasing for each cohort from generation to generation. Population extinction finally ensues, even when the one-time increase in mortality is a mere 2 percent. This simple model demonstrates how sensitive populations may be to minimal perturbation.

The model that Zubrow developed for an interaction between Neanderthals and modern humans showed itself to be even more sensitive. In this fully interactive growth model a population's mortality rate is governed by two factors: the mortality rate of the previous generation, and the mortality rate of the *other* population in the previous generation. The effect is as follows: a decline in mortality rate in one population increases the rate in the other population; a rise in mortality rate in one population causes a decrease in the second population. Even when the contribution to total mortality from competitive interaction with a second population is small (in the region of 2.5 percent), the effect of the interaction proves to be surprisingly large.

For a population that suffers a small mortality disadvantage in the first generation (as in the previous model), life expectancy declines in the first population and increases in the second, with the increase in life expectancy in the second population greater than the reduction in the first. In addition, the change in life expectancies in both populations is greatest among the younger cohorts, thus driving greater long-term consequences. The result is a multiplier effect, in which conditions change at an accelerating rate.

Zubrow has followed the outcome of many such population interactions, always with small percentages for initial disadvantage and contribution to mortality through competitive interaction, and always with the same result: rapid extinction for the initially disadvantaged population. When contemplating the possibilities of interaction between Neanderthal and modern human populations, he says, it is salutary to realize that a small advantage on the part of *Homo sapiens* (equivalent to the initial, one-time increase in mortality rate in their competitors in the models) could lead to rapid population replacement. If the local coexistence is between two bands of, say, 50 members each, then the loss of one Neanderthal individual could initiate the cascade to oblivion in less than a millennium. The loss of four individuals would speed the process, bringing extinction within 15 generations. That degree of demographic advantage might accrue either from occasional, limited violent encounters or from greater efficiency at harvesting limited resources. Zubrow's study does not rule out such violence, but shows, counterintuitively, that rapid population replacement can occur in its absence.

modern humans lived in the same area for at least a thousand years, perhaps longer. "This kind of thing makes the idea of killer Africans rampaging through archaic populations look ludicrous," says Chris Stringer. "With coexistence over this long a period, other explanations seem much more likely."

This exploration of some aspects of behavior relating to the origin of modern humans demonstrates clearly the complexity of patterns of evidence in human prehistory. It also reveals the emergence of one key human characteristic—that of ingenuity and innovation, particularly in the technological sphere. In the final two chapters we will look in greater depth at some of the less tangible aspects of behavior that characterize us as truly human: symbolism and language.

6

SYMBOLISM AND IMAGES

It is no simple matter to make a bead from ivory, especially if the only equipment available is the simplest of stone and bone tools. Cutting, grinding, piercing, and polishing operations are involved, at once time-consuming and delicate. A skilled worker can produce perhaps five beads a day, each measuring something less than a centimeter in diameter. "Beads have to be very important to you, if you are prepared to devote that much labor to them," observes Randall White, an archeologist at New York University.

In his office just off Washington Square, White has ivory tusks leaning against the wall, raw material for experimental investigation of prehistoric bead-making. Carefully stored in drawers in the same lab are scores of ancient ivory beads, products of human hands 35,000 years ago in the Aurignacian period—proof that technology

The Hall of Bulls, in Lascaux Cave, Dordogne, France.

had by that time been turned to matters beyond the utilitarian. "For the first time you get a strong sense of the modern human mind at work," observes White. "It's a reasonable assumption that the Aurignacian people [the first modern humans in western Europe] were using the beads in some form of body ornamentation, not in a trivial sense of simple decoration but as part of structured cultural expression."

Cultural anthropologists know that even technologically primitive people employ body ornamentation—whether of cloth, feathers, shells, paint, tattoos, incisions, or style of hair arrangement—as part of self and group identity. Idiosyncratic flourishes are common, of course, but beneath it all are rules of social structure, many perhaps vestiges of ancestral mores, but faithfully observed nonetheless. As Terrence Turner, an anthropologist at the University of Chicago, has observed: "The surface of the body . . . becomes the symbolic stage upon which the drama of socialization is enacted, and body adornment . . . becomes the language through which it is expressed. The adornment and public presentation of the body, however inconsequential or even frivolous a business it may appear to individuals, is for cultures a serious matter."

The stuff of archeological research is the tangible material objects that ancient people left behind. Body ornamentation, in the form of items such as beads and pendants, therefore provides fertile ground for archeological exploration of important social expression, that of identity and belonging. White has harvested an abundant crop of evidence for the appearance of body ornamentation in the western European prehistoric record, coincident with the first appearance there of modern humans, the Aurignacians. He sees in the archeological record an explosion of Aurignacian body ornamentation, which appears to have been complex conceptually, symbolically, technically, and logistically from the very beginning. No gradual transition from more primitive forms of symbolic expression; no steady development of the theme—instead, the sudden emergence of fully human behavior, presaged by nothing.

Art in a Human Context

As we shall see later in the chapter, the reality of this suddenness and its implications for the mode of evolution of fully modern humans is a matter of some disagreement among anthropologists. Nevertheless, when Aurignacian people started to use pendants and beads (presumably sewn in patterns onto clothing and arranged as bracelets and necklaces) as aspects of body ornamentation, they were opening up the chapter of human prehistory known as the art of the Ice Age, or Upper Paleolithic art. Between 35,000 and 10,000 years ago, people throughout western Europe (particularly France and Spain) were carving, engraving, and painting a wide range of images, sometimes in spectacular assemblages and often in deep, nearly inaccessible sites. The painted caves of Lascaux (in the Perigord region of France) and Altamira (in the Cantabrian region of northern Spain), the best-known examples of fine art, are both the product of peoples who lived less than 20,000 years ago.

Berkeley anthropologist Margaret Conkey says of these spectacular manifestations of Ice Age cultures: "Although the art at 20,000 years ago is not relevant to the *origin* of modern humans, it is relevant to what it *means to be* modern humans." During this early period, no matter where modern humans were—in Europe, in Africa, in Australia, perhaps in the Americas—art became a part of the human repertoire, an element of social, cultural context. "I suspect that it had something to do with negotiating social relationships," suggests Conkey. "This is how we see art and image making in modern ethnographic contexts."

Exquisitely carved horses: miniature, near-perfect renderings of "horseness" in ivory; bold, colorful images of bison, horses, and mammoths painted on cave walls, dwarfing the awed observer; curious compositions (a seal, a salmon, a snake, a tiny flower, some leaves, and other markings) etched into

These stylized human figures are possibly slaves roped together or warriors bearing staffs, trooping from right to left. Discovered by Louis and Mary Leakey in 1951, the paintings, at the site of Kundusi, Tanzania, may be 25,000 old.

a piece of reindeer antler; a preponderance of carnivore (lion and fox) teeth as pendants; an enigmatic ordering of images within caves; mysterious chimeras: part human, part beast. This was the art of the Ice Age, images from a culture—or, more properly, many cultures—long past. By now more than 200 deco-

rated caves have been discovered in Europe and more than 10,000 decorated objects (portable art, as it is often known). Such objects were sometimes pieces of art in themselves—a sculpted animal or an engraved stone tablet (perhaps used ritually)—and sometimes carefully decorated utilitarian objects, such as spear-throwers or hide-scrapers.

Art for Art's Sake?

Altamira was the first decorated cave to be discovered, in 1868. But more than three decades were to pass before prehistorians accepted the spectacular images as the products of a distant age. Even then, they were viewed as the simplistic products of simple minds. Recently, John Halverson of the University of California at Santa Cruz revived this hypothesis. "Artistic expression was a new-found power, an intellectual one as well as a motor skill, and repeated for its own sake," he proposes. "It is absurd to suppose that human consciousness as we know it appeared full-blown coincidentally with an anatomically *Homo sapiens sapiens* brain."

Depictions of horses, bison, and other animals appear as single individuals or sometimes as groups, but only rarely in anything that approaches a naturalistic setting. The images are accurate, but they are plucked from their context. To Halverson, this indicates that the artists of the Ice Age were simply painting or engraving fragments of their environment, in the absence of any mythological meaning. The artists were like automatons, producing images "unmediated by cognitive reflection," claims Halverson; the images are simple, depict no narrative, and show "nothing that can be attributed with any solid assurance to religious motivations." Human consciousness as we experience it was yet to be formed at this point in human prehistory, he argues. "At this early stage of mental development, percept and concept may have been undifferentiated."

Body Adornment: The Language of the Beads

Aurignacian sites, which contain implements made from fine flint blades and bone and ivory artifacts, are known in France, Germany, and Belgium. Aurignacian people lived in small foraging groups that hunted reindeer, woolly mammoth, wild horse, bison, and European red deer. Beads and pendants have also been found at such sites, fashioned from soft stone, shell, tooth, antler, and mammoth ivory. Many of these objects were perforated, presumably for suspension as pendants or for sewing onto cloth.

One of the features that distinguish Upper Paleolithic from Middle Paleolithic sites is the presence of material whose source is tens or even hundreds of kilometers distant. Such material, which may have been transported to the site by its occupants or, more likely, through exchange with other people, includes primarily the raw material for body ornamentation. These exotic objects clearly had value in a social if not an economic sense, and very probably indicated group identity and perhaps individual status within the group.

The notion of symbolism is even more cogent with the pendants fashioned from animal teeth, which seem to identify the wearers with their subsistence mode as hunters. Of the tooth-pendants recovered from Aurignacian sites in southwestern France, for instance, virtually all are from carnivores, particularly foxes. As British anthropologists Marilyn and Andrew Strathern have observed in their studies of New Guinea societies, self-decoration is rarely representational. When people wish to associate themselves with powerful forces in their cosmos—birds, for example—they do not make realistic masks but instead take parts of birds, such as bones and feathers, and wear these. The practice of evoking the whole through a part, termed metonymy, is well known to anthropologists.

Tooth-pendants from the French Aurignacian represent about one third of all objects of ornamentation. The remainder are mostly manufactured, labor-intensive beads and pendants. Randall White has studied such objects from three major Aurignacian sites in the Vézère Valley of southwestern France—Abri Blanchard, Castanet, and La Souquette. He identifies nine types of pendant, which represent at least five different production techniques: 1) animal teeth, perforated through the root; 2) circular or oval perforated beads made from bone, antler, or ivory; 3) elongated perforated beads made from ivory; 4) basket-shaped perforated beads made from stone or ivory; 5) engraved and perforated pendants made from bone or ivory; 6) tubular or disklike perforated beads made from stone; 7) intact but altered animal bones perforated for suspension; 8) fossils and worked bone, antler, or ivory objects circumscribed at one end for suspension; and 9) intact marine shells (fossil and actual) perforated for suspension.

Mammoths were rare in the French Aurignacian, and indeed in much of the Upper Paleolithic: there were no bones of these animals in the three caves in question. White speculates that the Vézère Valley may have been a center of exchange, with shells going east into what is now Germany and ivory following the opposite course. The production of beads in the three sites began with sticks of ivory that were never more than 10 centimeters long and from 0.45 to 1.4 centimeters in diameter. The sticks, cut from the softer outer layer of the mammoth tusk, must have been produced at the site of origin, because whole tusks are absent from the Vézère Valley locations.

The first stage involved segmenting the ivory sticks, which had been whittled to a cylindrical shape, by incisions around the circumference made every centimeter or two. Segments were snapped from the stick, forming the basic blanks, hundreds of which have been found at Aurignacian sites. Laminated layers of ivory were then removed asymmetrically, so that one end of the blank was thinner than the other. This crudely shaped blank somewhat resembles certain animal teeth, particularly the vestigial canines of the red deer and reindeer. White points out that much of Aurignacian art involves mimicry of natural forms; this resemblance was probably no accident. The upper section of the bead, at this stage about 0.2 centimeter thick, was crudely pierced at the junction with the thicker section, using a gouging action with a pointed stone implement. Subsequent grinding and polishing reduced the mass

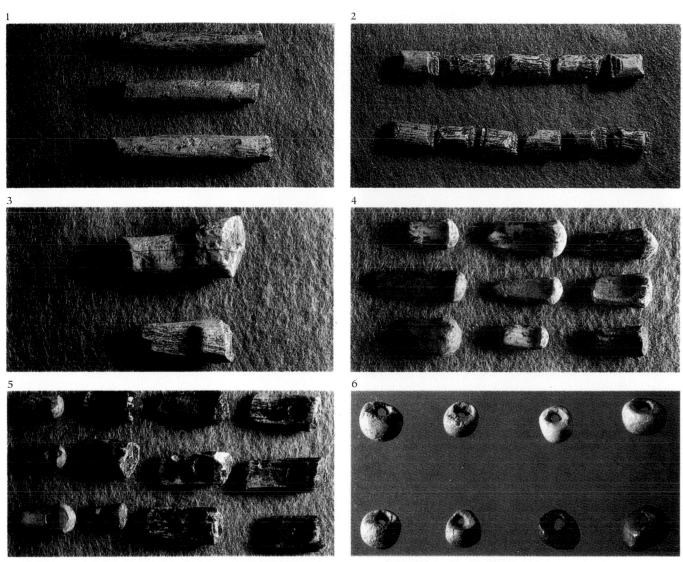

Bead production in the Aurignacian period in southwestern France involved several stages of manufacture. A rod of mammoth ivory (1) is incised around its circumference to form "blanks" for individual beads, 2). The blanks are shaped to yield a thick end and a thin end, the shape of a bulb, 3) and 4). The blank is pierced at the thin end, 5), and then ground and polished, yielding the final product, 6).

of the bead by half. Much of the stem disappeared, leaving a hole that gives the illusion of having been finely pierced.

Exactly how the beads were used is very much a matter of speculation. Analogy with Western society encourages the hypothesis that they may have been parts of necklaces or bracelets, although there is no unequivocal evidence in its support. Paired sets of beads have occasionally been found, which may suggest use as earrings or earplugs. Although discoveries from later in the Upper Paleolithic indicate that beads were sewn in profusion as patterns on clothing, no such direct evidence exists in the Aurignacian.

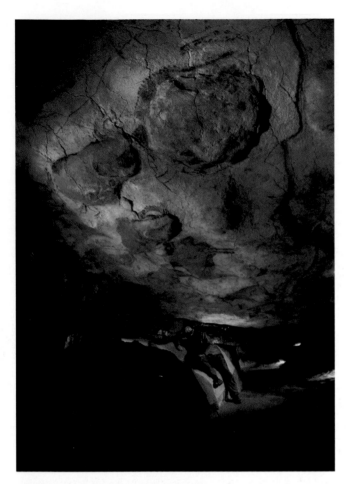

Altamira Cave, in Cantabria, northern Spain, was discovered in 1879 but was not accepted as genuinely prehistoric until 1902. Together with Lascaux in France, Altamira with its images of bison painted on a low ceiling is among the finest examples of Upper Paleolithic decorated caves.

In his strongly stated position, Halverson seems to ignore the human–animal chimeras, the red and black images of horses amid unrealistic cascades of dots, geometric signs of unknown identity or meaning, and many other examples that don't conform to the description of accurate representational art. Why, moreover, would the people have horses and bison on their minds (that is, in the paintings) when they had reindeer and ptarmigan in their stomachs? The notion that these images were "unmediated by cognitive reflection" is hardly tenable; some kind of selection and imagination was clearly at work.

Equally, how would a spectator of images from a distant (and now vanished) culture know whether religious motivation had operated or not? The image of shepherd and lamb is, to a Christian, rich with symbolism that invokes many aspects of the religion. Show the same image to a Yąnomamö Indian, however, and it signifies nothing beyond the simple representation of a man and an animal. Religious and mythological symbols are cryptic, speaking volumes to members of the culture concerned, but to an outsider they may be merely bafflingly strange images or apparently contextless everyday objects. Robert Layton, an anthropologist at the University of Durham, England, feels strongly that the images of Upper Paleolithic art were deeply meaningful in a mythological sense to the people of the time.

"It is hard to see how such a tightly organized expression of images that we see in the art could be without some kind of mythology behind it," he says. "But, alas, we will probably never know what it was." The desire to know and understand the origin of our own and other peoples' myths is very strong, says Layton. But, given the circuitous turns of expression and the importance of concealed symbols (complete and incomplete) that make up the telling of any myth, the chances of an outsider gaining more than a faint glimmer of understanding are small. "If you were to meet an Upper Paleolithic person and ask him to explain the myths embodied in the paintings, you'd probably finish up grabbing him by his fur lapels and shouting in frustration, demanding to know what he really meant by the incomprehensible and bizarre story he'd just told you," says Layton.

The cognitive ability to create images confers on humans the facility to manipulate objects in a manner that goes beyond language. Images that become symbols move in a medium of power. To many of the San people of southern Africa, for example, the eland symbolizes potency—in its living form, but more particularly when crafted as images by shamans. To the San, as we shall see in greater detail later, pictures of eland function as conduits to other realms. While the Upper Paleolithic people did not emphasize a single beast in the same way that the San reify eland, their art is dominated by horses and bison. Were these animals connected with the potencies of spirit worlds, as the eland is for the San?

Breuil's Hunting Magic, Leroi-Gourhan's Structuralism

Anthropologists have pursued the meaning of Upper Paleolithic art almost to the point of obsession, so tantalizing is it, at once eloquent and mute. The first major interpretation of Ice Age images beyond that of "art for art's sake" was that they represented a kind of hunting magic, a mythic system designed to ensure success in the food quest. The idea had been inspired by anthropologists' discovery at the turn of the century that many Australian Aboriginal paintings were part of magical and totemic rituals. The same might be true of Upper Paleolithic art, argued Salomon Reinach in 1903. Both peoples were hunter-gatherer societies, he noted; both produced paintings in which a few species were overrepresented by comparison with the existing environment. Therefore, Reinach reasoned, Upper Paleolithic people made paintings to ensure the increase of totemic and prey animals, just as the Australians were known to do. Anthropologists call this ethnographic analogy, a useful but by no means foolproof way of thinking.

The most famous name associated with the notion of hunting magic, however, was that of the Abbé Henri Breuil, who became strongly convinced by

Australian Aboriginal art was produced principally in a ritualistic context, often in connection with the spirit world; the example here, from the Northern Territory, Nourlangie, is of a six-fingered spirit.

Reinach's ideas and whose intellectual standing gave them great weight. In a distinguished career beginning in the first decade of the century and lasting almost 60 years, Breuil recorded, mapped, copied, and counted images in the caves throughout Europe. He also developed a chronology for the evolution of art during the Upper Paleolithic. Throughout, the notion of Paleolithic art as hunting magic was the ruling anthropological paradigm.

But the picture of hunters gathering in a dark and special place, there to invoke the spirits' aid for

The Abbé Henri Breuil, who established the study of decorated caves, seen here (pointing, at right) in Lascaux Cave shortly after it was discovered in 1940.

the next day's foray, is something of a simplistic caricature of the world of foraging people. To judge from a more complete reading of hunter-gatherers' lifeways, theirs is a world rich in meaning, animistic powers, and mythic explanation—of which their art is one element among many in coming to terms with a complex world. Hunting magic, pure and simple, is therefore likely to be an incomplete characterization of Paleolithic art.

When the Abbé Breuil died in 1961, an intellectual change was already under way, one that eventually ousted the all-encompassing earlier hypothesis. French archeologist André Leroi-Gourhan had begun his own investigations of the art, expecting to find "cultural chaos . . . works scattered over the walls in disorder by successive generations of hunters." The hunting-magic hypothesis had implied that images would be distributed haphazardly in the caves, and

Breuil had observed no order. Leroi-Gourhan, however, discovered that the images were not randomly positioned. "Indeed, consistency is one of the first facts that strikes the student of Paleolithic art," he said. "In painting, engraving, and sculpture, on rock walls or in ivory, reindeer antler, bone and stone, and in the most diverse styles, Paleolithic artists repeatedly depict the same inventory of animals in comparable attitudes. Once this unity is recognized, it only remains for the student to seek ways of arranging the art's temporal and spatial subdivisions in a systematic manner." But what did this order mean?

The fact that images painted in the caves were very often of animals not included in the painters' diet was a clear problem for the hunting-magic hypothesis. In many cases, discoveries of animal bones from living sites indicate that reindeer were important as food, yet reindeer images are few. The reverse was true for horses and bison. As Claude Lévi-Strauss once observed of art among San and Australian Aborigines, certain animals are depicted frequently not because they are "good to eat" but because they are "good to think." Leroi-Gourhan's challenge was to discover what Upper Paleolithic people had been thinking about when they decorated their caves. His answer was the division between maleness and femaleness.

In Leroi-Gourhan's scheme the horse image represented maleness and the bison femaleness; the stag and the ibex were also male, while the mammoth and the ox were female. His theory went beyond merely identifying associations of sex to include ideas concerning the distribution of these symbols throughout each cave. Deer, for instance, often appear in entranceways but are uncommon in main chambers, where horse, bison, and ox are predominant. Carnivores, rare images overall, mostly occur deep in the cave system. In some way, argued Leroi-Gourhan, the structure he saw in the caves reflected structure that was important to Upper Paleolithic people—perhaps the way society was organized, or the way they perceived the world around them.

Annette Laming-Emperaire, another French archeologist, agreed with Leroi-Gourhan that the images were ordered in the caves and that male-female duality was a part of that order. Unfortunately, in some cases the two prehistorians disagreed about which animals represented maleness and which femaleness—differences of opinion that "dealt a trump card to the critics of this new vision of Paleolithic art," said Laming-Emperaire. Toward the end of her life Laming-Emperaire began to develop the thought that some of the decorated caves represented origin myths, stories told through pictures; but she died in 1977, before carrying the idea very far.

The so-called Chinese horses in the Axial Gallery of Lascaux are among the most exquisitely crafted images in Upper Paleolithic art.

When Leroi-Gourhan died, in 1986, his ruling paradigm of all-embracing structure suffered the same fate as had Breuil's: it was overturned. "Archeologists began to look at the diversity in the art and away from monolithic explanations," explains Margaret Conkey. "There began to be a concentration on a diversity of meanings and a concern for the context of the art."

A Diversity of Interpretations

Leroi-Gourhan's death brought the second great era of the study of Paleolithic art to an end. Since then, no single figure has emerged to dominate the field, reflecting the newly emerging theme of diversity. For instance, Denis Vialou of l'Institut Paleontologie Humaine in Paris confirms that there is order in the distribution of images within caves, but not the kind of global order that Leroi-Gourhan talked about. "Each cave should be viewed as a separate expression," says Vialou. Meanwhile, Henri Delporte, of the Musée des Antiquités Nationales, near Paris, is comparing wall art and portable art, looking for different kinds of patterns. Conkey stresses the social context of the production of images. Randall White is analyzing the technology behind the image-making, as are French archeologists Jean Clottes and Michel Lorblanchet; and so on.

Drawing attention to the modern-day bias through which, inevitably, we perceive the art of the Ice Age, Conkey says, "Perhaps we have closed off certain lines of enquiry simply by attaching the label 'art' to Upper Paleolithic image-making: this presupposes an aesthetic element and imposes a Western evaluation of the activity." Indeed, Leroi-Gourhan once described the great painted caves of the middle of the Ice Age as "the origins of Western art"— clearly an erroneous evaluation, partly because this form of image-making came to an end about 10,000 years ago and partly because some of the techniques seen in the Ice Age images (such as perspective) had to be reinvented in the Renaissance. In an attempt to get beyond this kind of bias, Conkey suggests that "rather than ask what the images *mean*, we should look to see what made them meaning*ful*"—a slight but important twist, she maintains.

The term Ice Age art is often taken to imply a kind of uniformity of expression. This is simply incorrect; there was a great diversity of art, both in space and time. In the previous chapter, the four cultural eras of the Ice Age—the Aurignacian, Gravettian, Solutrean, and Magdalenian—were described, usually delimited by changes in material technology. Change in image-making occurs through these eras too, both in emphasis and in medium.

As noted at the beginning of this chapter, the first era of the Upper Paleolithic, the Aurignacian, is notable for the sudden burst of body ornamentation and carved and engraved objects, while painted caves are virtually absent. Not until the Magdalenian era (18,000 to 10,000 years ago) did cave painting become really important. In fact, 88 percent of all Ice Age images come from this late era of Lascaux and Altamira, leading some scholars to challenge the notion of an explosion of artistic expression from the beginning of the Upper Paleolithic, a notion we will return to later.

Often unmentioned in discussions of Upper Paleolithic cultures is music-making, so much in the thrall of the magnificent painted images are we. From the Aurignacian to the Magdalenian, Upper Paleolithic people made flutes (from bird, reindeer, and bear bones) and, presumably, other instruments. Music is very much a part of making and using images for the San and the Australian Aborigines; perhaps the same was true for the Upper Paleolithic hunter-gatherers of Europe. Gradually, a picture may be built up of Upper Paleolithic people as skilled hunter-gatherers with a rich mythology, who engaged in widespread social and political alliances enhanced by trade and exchange.

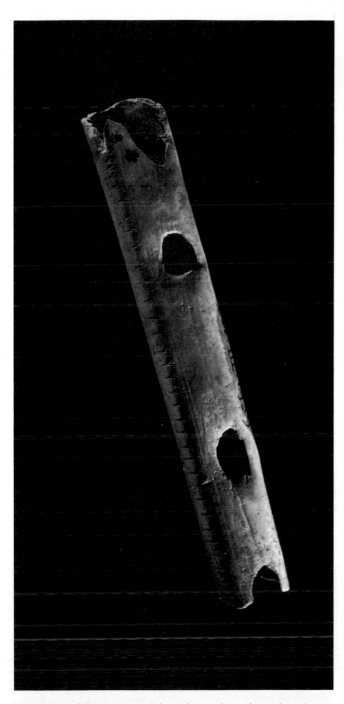

Examples of flutes, carved from bone, have been found from the earliest Upper Paleolithic sites. This one is made of bird bone and is 25,000 years old.

Points of Aggregation

Margaret Conkey has suggested, for instance, that the cave of Altamira might have been an aggregation site: a place upon which neighboring bands converged during the fall. The reason for the aggregation would have been social and political as much as economic—just as among hunter-gatherers of historical times. For example, the !Kung San people of the Kalahari in southern Africa spend much of the year as members of small foraging bands composed of about 25 individuals, a collection of nuclear families. During the dry season, however, several bands aggregate around permanent waterholes, when more than a hundred individuals will live and forage in relative proximity. Although the aggregation is explained as an economic necessity—secure access to water   the social interaction among the groups is intense, with marriages arranged and performed, political alliances strengthened and modified.

The G/wi San, neighbors of the !Kung who live in even more ecologically marginal territory, also spend much of the year as small, mobile bands, occasionally aggregating for ostensibly economic reasons. Aggregation occurs during the rainy season, when temporary waterholes form, but the most important consequence, again, is social and political interaction. Such patterns are repeated among all foraging people. The same may have been true of the foragers of the Upper Paleolithic. Conkey has speculated that the arrangement of bison and other animal images on the painted ceiling of Altamira may reflect the perceived social interaction of different bands, dispersed but allied. Almost two dozen polychrome images of bison form the main structure of the painted ceiling, arranged principally around the periphery. These, suggests Conkey, may represent the different groups that aggregated at the site. Other images are present, including two horses, a wolf, three boars, and three female deer.

The archeological evidence that encouraged the speculation about Altamira as an aggregation site

Pigments of the Imagination

Of all the artifacts left behind by prehistoric people, painted images on rock shelters and cave walls are among the most evocative. They speak to the Western mind of religion and ritual, something deeply important to an ancient mythology. Yet the images evade ready interpretation. In recent years chemistry and physics have come to the archeologists' aid in establishing some of the context of the paintings.

The first direct dating of European cave paintings, for instance, using the accelerator mass spectrometry radiocarbon technique on charcoal particles in the pigment, was reported by a French research team in 1990. Since then some of the most important cave art sites have been examined using the same approach, although it will be many years before a significant proportion of the paintings has been dated. Precise dating of painted images may establish a chronology of Upper Paleolithic people and their work; moreover, it will eventually allow an insight into stylistic variability, which may be related to cultural expression.

Until direct dating of images was possible, indirect methods were used, such as, where feasible, the dating of material in occupation layers thought to be associated with the paintings. The French prehistorians Breuil and Leroi-Gourhan each developed chronologies of the painted caves based on their interpretation of the change of painting style through time. The uncertainty of this ap-

These bison images in the Salon Noir of Niaux Cave, Ariège, France, were done by Late Magdalenian artists, about 12,000 years ago.

proach was revealed in 1992 when a team of French and Spanish scientists published radiocarbon dates on stylistically similar images of bison from the caves of Altamira and El Castillo, in Spain; and for a stylistically different bison in the cave of Niaux, in the French Pyrenees. The dates were, respectively, 14,000, 12,990, and 12,890 years. Clearly, age and style do not always coincide.

The analysis of pigments has gone beyond dating. Jean Clottes, the French government's director of prehistoric research in the Pyrenees, in collaboration with Michel Menu and Philippe Walter of the research laboratory at the Louvre Museum, Paris, gained some insight into the technology of Ice Age painting.

Among their discoveries is that while the technology of paint production may change through time, style may persist.

The cave most studied so far is Niaux, in the limestone massif of the Vicdessos Valley, which is counted among the half-dozen finest examples of Ice Age art in Europe. The painted images in the virtually horizontal galleries, which extend more than 2 kilometers, were discovered in 1906 and have since been the subject of considerable scrutiny. The most decorated section of the cave system, the Salon Noir, is a side chamber—a soaring vault far from daylight where mainly black images of bison, horses, deer, and ibexes decorate the walls, arranged in panels and giving the impression of foresight and deliberation in their execution.

The black pigment used at Niaux has long been thought to include a mixture of charcoal and manganese dioxide. Clottes and his colleagues recently applied modern analytical techniques to 59 minute samples of black and red pigment from many images throughout the cave, and from the adjoining cave of Réseau Clastre. Using a scanning electron microscope attached to an X-ray detector (to identify specific minerals) and proton-induced X-ray emission (to determine the complete elemental composition), the researchers obtained a detailed profile of the chemical components of the paint. Although all are naturally occurring and local, their combination in the paints is clearly man-made.

The painting materials—pigments and mineral extenders—were carefully selected by Upper Paleolithic people and ground to within 5 to 10 micrometers to produce a specific mix. The black pigment, as had been suspected, was charcoal and manganese dioxide. But the real interest was in the extenders, of which there seemed to be four distinct recipes, which the researchers number one through four. Extenders help to bring out the color of the pigment and, as their name implies, add bulk to the paint without diluting the color. The four recipes for extenders used at Niaux were talc; a mixture of baryte and potassium feldspar; potassium feldspar alone; and potassium feldspar mixed with an excess of biotite. Clottes and his colleagues experimented with some of these extenders and found them to be extremely effective.

In the Salon Noir most of the paintings were made with a mixture of charcoal and manganese dioxide, but it was unlike any other such combination in the cave. The particles of charcoal were unusually large and were consistent with a charcoal stick being used to sketch an outline. On top of this initial sketch was a paint composed of manganese dioxide and one type of extender, potassium feldspar and biotite (recipe four). By examining the chemical profile of minor constituents in the paint, the researchers discovered that in one panel that was examined in some detail, several different "paint pots" must have been used. While the recipe for the extender was identical throughout the panel, which consists of three bison, two different paint mixes had been employed. Clottes and his colleagues speculate that the panel may have been painted in more than one session, or perhaps worked on by different painters who followed the same recipe but obtained their components from different locations.

The overwhelming majority of paintings in the Salon Noir included recipe four, but the other three types of extender are present, too. Clottes believes that the fact that most of the paintings used recipe four implies a coherence in their execution, possibly in time but certainly in cultural context. He suggests that the ritual that undoubtedly attended the production and use of images deep underground may have extended to the manufacture of the paints themselves.

Preliminary dating of some of the images throughout the cave shows that different extenders were used at different times: those used between 14,000 and 13,000 years ago included feldspar, while later recipes, up to 12,000 years ago, favored large quantities of biotite. Whether practical issues, such as ready availability, or ritual concerns governed this shift is unknown. It is surely significant, however, that the style of the paintings done through that time remains remarkably constant.

The principal conclusion concerning the Salon Noir is that it should be regarded as an important sanctuary: the outline sketch of the panels, followed by their full execution, speaks of a premeditation absent from most of the other images, often deeper in the cave. Such premeditation is consistent with a ritual ceremony of considerable importance, suggests Clottes. Paintings made without a preliminary sketch could have been done very quickly, as the French prehistorian Michel Lorblanchet has demonstrated by experimental replication of some paintings. Produced during a single visit to a deep part of a cave, perhaps never to be revisited, such images may nevertheless have played an important role in Upper Paleolithic mythology. Chemical analysis may help define the parameters of how and when images were painted, but even these powerful techniques fail to answer the most interesting question of all: what did the images mean to Upper Paleolithic people?

concerns the engraved objects found there. Many different designs are found on ivory and stone objects in the Cantabrian region in the Magdalenian period: chevrons, lunate structures, nested curves, and so on. Among these many design elements 15 basic forms have been identified, each of which tends to be geographically restricted, suggesting local styles. At Altamira, however, many of the different local styles are found together; different groups may have brought these objects to Altamira individually, but the notion of aggregation is plausible. The fall abundance of red deer and limpets would have been the

Geometric patterns, such as grids, circles, and dots, were once interpreted as depictions of hunting equipment, such as traps. Such images are, however, consistent with certain simple hallucinations that occur in trance states, and may indicate that some Upper Paleolithic art is shamanistic in character. This example is from Castillo Cave, northern Spain.

ostensible reason, but social interaction the ultimate purpose.

So far, no such archeological evidence has been uncovered at Lascaux to suggest that it, too, may have been a major aggregation site; there is no apparent ecological reason, moreover, for gathering there at any particular season (when, for instance, large game or fish were plentiful). Nevertheless, the special assembly of images in the cave could have provided reason enough for the spot to have assumed a status justifying periodic aggregation. For Australian Aborigines, geological landmarks such as mountains and caves often became socially important not for ecological reasons but because of their links to mythic events relating to activities of ancient spirits.

In his studies of many of the sites in northern Spain, Robert Layton has perceived a pattern of nonrandom distribution. He concludes that each of the major ones, including Altamira, appears to be at the center of a scattering of smaller sites. "The more elaborate sites seem to be a focus of some kind," he says. According to his calculation, each of the major decorated caves has a "sphere of influence" about 30 kilometers in diameter—perhaps a spatial pattern of political power, the optimum distance over which alliances could readily be maintained. No such patterning has yet been discerned among the cave sites of France.

Indications of Shamanism

One of the great enigmas of Upper Paleolithic art is the presence of curious geometric patterns—grids, lines of dots, nested curves, chevrons, rectangles, and so on—scattered among the realistic images of animals. Definitely not a normal feature of most Western art, these "signs," as they are called, are a constant component of Upper Paleolithic art. The Abbé

Breuil proposed that these geometric patterns represented hunting paraphernalia: traps, snares, even weapons. This fitted his interpretation of the art as hunting magic. Leroi-Gourhan included them in his notion of a structural duality in the art. Dots and strokes were male signs, he said, while ovals, triangles, and quadrangles were female signs.

Recently a South African archeologist, David Lewis-Williams, has suggested that neither interpretation is correct. Rather, he says, the images are plucked from a mind in the state of hallucination: a sure sign of shamanistic art. His argument is based on a study of San art, from southern Africa, and on a neuropsychological model that may be basic to much human image-making in hunter-gatherer societies, including those of the Upper Paleolithic. Others have suggested previously that Upper Paleolithic art might have shamanistic elements, but the neuropsychological model that Lewis-Williams developed with his colleague Thomas Dowson elaborates the idea much further. It should be said that the model is controversial among archeologists, although its relevance to current discussions was acknowledged when Lewis-Williams was invited to make a presentation at the half-century anniversary of the discovery of Lascaux in 1990.

Lewis-Williams began his work with an attempt to understand the prehistoric rock art of southern Africa at Apollo-11 Cave in Namibia, some of which goes back as far as 27,000 years, a full 10 millennia before Lascaux. The African art is much more schematic and contains more human figures than the European Ice Age art, but it includes the same kinds of geometric signs. San art is a recent part of this body of images in southern Africa, and for a long time it was considered to be principally narrative and decorative, with a small element of mythological depiction. After some false starts and frustrations, Lewis-Williams turned to ethnography for inspiration, specifically to accounts of the Kalahari San. The last

of the images here were painted about 120 years ago, at the fringes of recent history.

Prominent among San images, as mentioned earlier, was the eland, just as it was over much of southern Africa. Immersion in Kalahari ethnography and anthropological theory produced the notion that the eland was a symbol with multiple meanings. It turned up in all kinds of contexts, in rituals of the girl's puberty, the boy's first kill, marriage, rainmaking, healing—everywhere. But it became clear that the eland image was most prominent in shamanistic, as opposed to narrative and decorative, compositions. Shamans, viewed as possessing special powers for making contact with the spirit world, perform an important role in maintaining a society's mythology. Among the San this involves self-induced trance during ritualized ceremonies, often followed by the painting of images. As a result San art is dominated by shamanism.

In his studies on San art Lewis-Williams made contact with an old woman who lived in the Tsolo district of the Transkei. The daughter of a shaman, she claimed intimate knowledge of the now vanished shamanistic rituals. She explained that shamans may induce trance in themselves by various techniques, including drugs and hyperventilation, but almost always in the context of rhythmic singing, dancing, and clapping by groups of women. As trance deepens, the shamans begin to tremble, their arms and bodies vigorously vibrating. They may also bend over, as if in pain, and need the support of a staff. They are "dying," visiting the spirit world beyond this one; they are also hallucinating. Cuts may be made on the forehead and neck of an eland and the blood used to infuse potency into a person by rubbing it into cuts on his neck and throat. Later, the shamans paint images of the hallucinatory experience, often using some of the blood and producing a record of the contact with the spirit world in that particular context. The old woman told Lewis-Williams that power exuded

This example of San art, from Bamboo Hollow, Natal Drakensburg, southern Africa, was once interpreted as a display of "hunting magic" but is now recognized as shamanistic in origin. The animal is not an intended victim of the hunt, but is some kind of "rain animal," and the figures are not hunters but shamans in a state of trance.

from the paintings and could be acquired by placing one's hand on them.

Lewis-Williams searched the scientific literature for reports of the experimental study of visual experience during hallucination. He found that there had been interest in the topic in the late nineteenth and early twentieth centuries, but systematic research began with the work of the German psychologist Heinrich Klüver in the early 1940s. He and half a dozen researchers who came to the investigation later discovered that an experimental subject in an altered state of consciousness went through a rather regular set of visual experiences, not random, anarchic sensations as had been thought. Researchers used many different methods to induce altered states of consciousness, including electrical stimulation, flickering light, psychedelic drugs, auditory deprivation, and hyperventilation.

The research showed that people in early trance states initially "see" so-called entoptic phenomena: shimmering, incandescent, moving patterns produced within the visual system. (Entoptic, from the Greek, means "within vision.") These entoptic images take geometric form—grids, parallel lines, zigzags, dots, spirals, nested sets of curves, and filigrees. In deeper trance states, these geometric images may be construed as real, depending on the state of mind and cultural background of the individual. As one researcher put it: "Thus the same ambiguous round shape on initial perceptual representation can be 'illusioned' into an orange (if the subject is hungry), a breast (if he is in a state of heightened sexual drive), a cup of water (if he is thirsty), or an anarchist's bomb (if he is hostile or fearful)." One construal common among the San is to see nested curves as honeycomb, a substance important to these people.

Moving into a yet deeper stage of trance is often accompanied, according to laboratory reports, by the experience of a vortex or rotating tunnel that seems to surround the subject. The external world is progressively excluded, and the inner world grows more florid. Iconic images may appear on the walls of the vortex, often imposed on a lattice of squares. Frequently there is a mixture of iconic and geometric forms. Experienced shamans are able to plunge rapidly into deep trance, where they manipulate the imagery according to the nature of the ritual upon which they are about to embark. They view their hallucinatory experience, however, as visions of a world they have come briefly to inhabit, not a world of their own making: a spirit world they are privileged to visit.

Hallucination in this deepest of the trance stages can become quite bizarre, with human and animal forms combining, forming chimeras (or therianthropes, from the Greek roots for *beast* and *human being*, as they are also known). One early

experimenter reported the transformation of a human head into a cat's head. Another described his experience this way: "I thought I was a fox, and instantly I was transformed into that animal. I could distinctly feel myself a fox, could see my long ears and bushy tail, and by a sort of introversion felt that my complete anatomy was that of a fox."

From these and other details of the neuropsychological literature, Lewis-Williams and Dowson developed what they call their three-stage neuropsychological model, which corresponds to the three increasingly deep trance stages. If shamanistic art derived ultimately from the three levels of hallucination described by the model, it should contain images corresponding to these stages. For instance, there should be a certain number of geometric signs (entoptic images), some representational images (construals), and human–animal chimeras (therianthropes). This is precisely what is seen in San art, which is known to be shamanistic.

"The neuropsychological model orders and fits Upper Paleolithic art as well as it does San art," says Lewis-Williams. "The 'fit' is by no means simple; it is in fact highly complex, and this increases confidence in the conclusion that it does not result from chance." For instance, all six entoptic forms are to be found in Upper Paleolithic art, although this does not account for all geometric forms in the caves. The presence of those linked to altered states of consciousness was, however, seen as an apparent confirmation of stage one hallucination and associated images.

Stage two, the construal of entoptic forms, proved more difficult to identify. Nevertheless, Lewis-Williams and Dowson suggest that the exaggerated, curved ibex horns with a zigzag outer margin in the Pyrenean cave of Niaux might qualify. So, too, might similarly exaggerated horns of ibex on a stone lamp from the cave of La Mouthe in the Dordogne. The South African researchers also cite paintings and engravings of mammoths in Rouffignac, also in the

Dordogne, in which the tusks of various animals appear as nested curves. These and other suggestions are tentative, admit Lewis-Williams and Dowson, but they "expect examples as convincing as the San ones to come to light."

Stage three, the most complex of all, provides stronger evidence. Looking for combinations of iconic and geometric images, Lewis-Williams and Dowson cite the famous horses of Peche Merle, where red and black dots were placed within the black outlines of the animals. With the more enigmatic images of therianthropes, there are strong parallels between San and Upper Paleolithic art. Often in the past such im-

ages were interpreted as hunters or shamans wearing masks; this does not explain, of course, the common occurrence of therianthropes with hooves as feet. There was once even the suggestion that therianthropes were the product of a "primitive mentality" that had "failed to establish definitive boundaries between humans and animals."

This last suggestion aside, Lewis-Williams and Dowson acknowledge that during certain rituals shamans may have used animal adornments, but argue that the overall nature of therianthropic images is strongly consistent with stage three trance. This holds as persuasively for paintings and engravings in Upper Paleolithic art as it does for San art.

The so-called Sorcerer, from the Trois Frères Cave, France, is an example of a part-human/part-animal figure, a therianthrope, which may be an important clue to the nature of some Upper Paleolithic art.

The Mystery of Chimeras

Therianthropes represent a small but arresting proportion of Upper Paleolithic images. The most famous example is the so-called sorcerer in the cave of Les Trois Frères in the French Pyrenees. Deep underground, the sorcerer dominates the cramped cavern. Denis Vialou, of l'Institut Paleontologie Humaine, Paris, who studied the cave in detail, describes the image like this: "The body is uncertain, but is some kind of large animal. The hind legs are human, until above the knees. The tail is some kind of canid, a wolf or a fox. The front legs are abnormal, with human-like hands. The face is a bird's face, odd, with deer's antlers." Unusually for Upper Paleolithic images, the sorcerer is gazing directly out of the wall, a full-face stare that transfixes the spectator.

Below the sorcerer are several heavily engraved panels, a riot of animal figures with no apparent order. In the midst of all this is another human–animal figure, again with human hind legs. Human hind legs on animals are common in Upper Paleolithic art, as are hooves on otherwise human figures. This therianthrope is standing upright, with a bison's body, head, and horns but a somewhat human face.

The front legs are odd, in the same way as the sorcerer's forelimbs. This individual is holding something that might be a bow or a musical instrument. "Directly in front of this image is an animal," explains Vialou. "It has reindeer hind legs and rear end, showing female sex prominently displayed, the only such display in Upper Paleolithic art. The rest of the body is bison, the head turned, looking back over its shoulder at the first individual. Something special is going on between these individuals, I'm sure of that."

Something similar is to be seen in Lascaux. The very first beast in the stampede in the Hall of Bulls is an enigma. Known as the unicorn—although it has two very straight horns—this beast has a swollen body on thick limbs, with the head of no known animal. There are six circular markings on the body, and the partial outline of a horse. Look at the head again, squint, and the profile snaps into that of a bearded man. It is a curious image, one that Lewis-Williams and Dowson believe fits very well into the kind of therianthropes hallucinated during stage three trance.

In arguing that Upper Paleolithic art conforms to the three-stage neuropsychological model and is therefore shamanistic, the South African researchers

This enigmatic figure in the Hall of Bulls in the cave of Lascaux is often known as the unicorn, despite clearly having two horns. Part-human/part-animal, the figure is an example of a therianthrope, which is found in Upper Paleolithic art and may be a clue to a shamanistic element in the art.

do not claim to have explained the meaning of the images. "Meaning is always culturally bound," say Lewis-Williams and Dowson. "What we are pointing to are neurological mechanisms that underlie shamanistic art, wherever it is produced. How people construe and manipulate entoptic images, and what kind of iconic images they depict, all this will be influenced by the cultural context." It is like working with a given palette of paints, from which any desired image may be constructed. Meaning of the images remains elusive, but a knowledge that the art is shamanistic—if indeed it is—at least offers a foundation from which to analyze it.

If Upper Paleolithic art is shamanistic then, suggest Lewis-Williams and Dowson, a clue might be had to the origins of image making—the notion that two-dimensional lines can represent three-dimensional objects. Prehistorians and psychologists have speculated over this issue for decades, wondering about the mental processes that would be required to provide the key innovation. One idea, recently revived, is that the production of representational images grew out of nonrepresentational marks as a progressive process, a maturing of cognitive abilities and insights. The suggestion that the art is shamanistic makes this unnecessary.

Some of the studies on hallucination report that the images, both geometric and iconic, often appear to exist on surfaces, as if projected onto the wall or ceiling: "pictures painted before your imagination," commented one observer. In a more naturalistic context, shamans often perceive their hallucinations as if they were on rock surfaces, which are viewed as an interface or passageway between the real world and the spirit world. "They see the images as having been put there by the spirits, and in painting them, the shamans say they are simply touching and marking what already exists," explains Lewis-Williams. "The first depictions were therefore not representational images in the way you or I think of them. They were fixed mental images of another world."

Magnitude of Change Debated

No doubt anthropologists will continue to probe for meaning in Upper Paleolithic art, seeking to understand the cultural complexity of an ancient lifeway. Clearly the product of the modern human mind, yet our minds cannot yet encompass it. While the search for meaning goes on, descriptions of the overall pattern are being challenged—specifically, the assertion that with the beginning of the Upper Paleolithic there was an explosion of artistic and symbolic expression. The pace of its appearance—sudden or gradual—is closely tied to ideas about the origin of modern humans. If, as proponents of Multiregional Evolution believe, modern *Homo sapiens* emerged gradually from archaic *sapiens*, then the expression of the modern human mind in an artistic form might be expected also to be gradual. If, however, the Out of Africa model is correct and modern humans evolved as an abrupt speciation event, then artistic abilities might well have emerged abruptly, too; and as these people arrived in Europe, they would bring their fully developed artistic powers with them. (In Africa, incidentally, there is little evidence of artistic or symbolic expression earlier than the Later Stone Age, except perhaps the production of some ostrich eggshell beads.)

Two anthropologists at the University of Pennsylvania, Philip Chase and Harold Dibble, recently surveyed the evidence for artistic and symbolic expression in the Middle to Upper Paleolithic periods, with the explicit purpose of determining the mode of the transition. Their conclusion was quite firm: "The most striking difference between the Middle and Upper Paleolithic is the contrast between the rich and highly developed art found in the latter period and the almost complete lack of it in the former." Chase and Dibble acknowledge that Middle Paleolithic people undoubtedly cared for each other to a much greater extent than is typical among other primate species, as is evidenced by "the survival of aged individuals and of individuals suffering from moderately severe physical handicaps." In addition, "the evidence of burials implies the presence of strong emotional bonds, so that even dead members of one's group were afforded treatment not found among nonhominid primates." But no fully expressed symbolism or image-making is demonstrable.

There is some evidence of image-making earlier than the Upper Paleolithic, but it is very limited: for instance, a fragment of bone marked with a zigzag motif from the Bacho Kiro site in Bulgaria, somewhat earlier than 35,000 years ago; a carved mammoth tooth, worn smooth with use and marked with red ocher, from the 50,000-year-old site of Tata, Hungary. Oldest of all is an ox rib engraved with a series of double arcs, from the French site of Peche de l'Azé, some 300,000 years old. Ocher has been found at several ancient living sites, including the campsite

of Terra Amata in southern France, dated to about 250,000 years ago. Nevertheless, argue Chase and Dibble, none of this betrays modern human symbolism at work, merely weak glimmerings. And many of the supposed elements of evidence of Neanderthal mythology, such as the Cult of Skulls, they call products of the overinterpretation of equivocal evidence by eager investigators.

The Cult of Skulls refers to a supposed practice of cannibalism in which an individual's brain is removed and the cranium placed in a ritual context. The most famous example of this is the Monte Circeo I cranium, from the Grotta Guarttari, Italy. Discovered in 1939, the Neanderthal skull was said to have been broken open at its base for the removal of brain tissue, placed upside down on the cave floor, surrounded by a circle of stones, and honored by ritual offerings of deer bones placed around it. This and several similar discoveries have been adduced as evi-

Detail of an ox rib, probably 300,000 years old, from Peche de l'Azé, France, bearing a series of engraved arches; if the arches truly are engraved, the piece would be a candidate for one of the oldest known engravings.

dence of Neanderthals' highly developed mythology and sense of the abstract. In recent studies, however, several researchers working independently have shown that Monte Circeo is not what has been assumed. For instance, Tim White of the University of California, Berkeley and Nicholas Toth of Indiana University demonstrated that there are no signs that the skull was deliberately severed from a body and broken at its base for access to the brain. Such activities, known from ethnographic example, leave characteristic cut marks in the bone; these are absent from Monte Circeo I. Mary Stiner of the University of New Mexico has shown that the collection of deer bones in the cave is characteristic of assemblages that accumulate in hyena dens. Giacomo Giacobini of the University of Turin supports Steiner's conclusion and argues that the circle of stones—often called the "crown of stones" in the literature—was more in the eyes of the beholders than a reality.

Careful studies of this kind are important in an accurate assessment of the cognitive abilities—particularly in abstraction and symbolism—of archaic humans. For Chase and Dibble such studies contribute to the conclusion that modern human image-making and symbolism arrived with modern humans. But not everyone agrees with this position. John Lindly and Geoffrey Clark, for instance, of Arizona State University, strongly object. "We are concerned that Chase and Dibble's conclusions might be taken by anthropologists inclined to see marked discontinuity across the Middle/Upper Paleolithic transition as further 'proof' of a major difference between these two periods and, consequently, considerable evolutionary 'distance' between archaic *Homo sapiens* and morphologically modern humans," they state in a response. In their examination of the archeological record, Lindly and Clark see no evidence of the appearance of symbolism coincident with the first appearance of anatomically modern humans (nor does anyone else); they argue that the Middle to Upper Paleolithic tran-

sition, as far as artistic expression is concerned, is a gradual, not a punctuational event; and they note that the complexity of artistic expression in the Upper Paleolithic increases with time, the Magdalenian being more developed than the Aurignacian.

Randall White disputes Lindly and Chase's contention that the Aurignacian is somehow poorer artistically than later periods in the Upper Paleolithic. "I have been struggling to understand the rich body of Aurignacian and Gravettian evidence, especially body ornamentation, from Western, Central, and Eastern Europe," he says; "the quantity of material is staggering." Paul Mellars of Cambridge University is equally unimpressed with this point of Lindly and Clark's. "I have never really understood the argument that the significance of the symbolic and technological 'explosion' at the start of the Upper Paleolithic is in some way diminished by the evidence of further increases in 'cultural complexity' during the later stages of the Upper Paleolithic," he notes. For Mellars, such changes are to be expected as population density increases and cultural evolution continues: "To argue that this evidence for later Upper Paleolithic cultural 'intensification' rules out the significance of the far more radical innovations in behavior at the *start* of the Upper Paleolithic would seem akin to dismissing the significance of the Neolithic Revolution on the grounds that things became even more complicated during the Bronze Age."

The Challenge of Interpreting Symbolic Expression

One of the problems archeologists face in seeking the origins of image-making and symbolism is that they are restricted to analyzing tangible, preserved products that our modern eyes recognize as complex. The simplest of objects—a naturally rounded stone, for instance—could be the centerpiece of a highly complex ritual; yet we might not recognize it or be able to reconstruct it. Equally, images created on material that vanishes from the record—on bark or in sand, for example—will never be seen by researchers' eyes. As we have earlier observed, however, to state that absence of evidence of image-making and symbolism is not necessarily evidence of its absence is special pleading of the weakest kind. For various archeologists to examine the same body of evidence and come to diametrically opposite conclusions is a problem of a different kind, one as much in the realm of the sociology of science as that of its methodology.

However it came about, we know that image-making and symbolism did become an important part of modern human culture; and it continues to be so. Our concluding chapter will explore the emergence of a closely allied phenomenon, that of language and consciousness. All are linked together in an intimate cognitive nexus.

7

LANGUAGE AND MODERN HUMAN ORIGINS

According to most estimates, some 5000 human languages exist today. Princeton anthropologist Clifford Geertz has observed that "one of the most significant facts about us may finally be that we all begin with the natural equipment to live a thousand kinds of life but end in the end having lived only one." He might have said 5000 kinds—or more accurately, a virtual infinity, for language is the principal tool for building human culture, and in theory there is a potential infinity of languages. This most powerful of communication channels in the biological realm fashions myriad worlds effectively divided by cultural barriers. Each culture is defined by a mythology, traditions, mores, shared morals, history—effectively, a shared consciousness. And shared language is the architect of it all, individually and collectively.

Our capacity to learn, our extreme plasticity in the face of new environments and new challenges, is often noted as an important

Crania sometimes become filled with fine mineral, which, during the fossilization process, sometimes forms a natural cast of the shape of the brain as it impresses itself on the interior surface of the skull. Such natural casts are a guide to brain reorganization during hominid evolution.

trait in *Homo sapiens*. But, as Geertz points out, what is unique is our dependence upon learning, not just our facility for it. "Beavers build dams, birds build nests, bees locate food, baboons organize social groups, and mice mate on the basis of forms of learning that rest predominantly on the instructions encoded in their genes," says Geertz. "But men build dams or shelters, locate food, organize their social groups, or find sexual partners under the guidance of instructions encoded in flow charts and blueprints, hunting lore, moral systems and aesthetic judgments: conceptual structures molding formless talents."

Language—without doubt one of the most important elements in bringing humans to the exquisite point of formless talent—presents many challenges to anthropologists. Some are easily formulated, others less so. When did spoken language attain the level we now experience? Were Neanderthals, for instance, less language competent than modern humans? In what function did natural selection favor so complex a talent? As we shall see, the obvious answer—that of more efficient communication among individuals—may not be correct; indeed, some scholars suggest that language capability was not selected for at all, but emerged as the fortuitous by-product of an enlarged brain. How would these and other issues relating to the evolution of a fully propositional language impress themselves upon the prehistoric record? (The term *propositional language*, formulated by René Descartes, means the "faculty of arranging together different words, and composing a discourse from them," the basis of human rationality.) However, central to what it is to be human within a human culture, a less tangible activity than language, in terms of the archeological record, can hardly be imagined.

Much of the reams of neurobiological experimentation and speculation about the nature of human language are beyond the scope and goals of this chapter. Here, the principal focus will be on the two questions that relate most pertinently to modern

human origins: First, when did modern human language arise? And second, what was its function in an evolutionary context?

Inevitably, discussion of these issues with respect to the evolution of language evokes inquiries made in earlier chapters. Did it arise rather suddenly—a punctuational leap at the very threshold of modern human existence, perhaps at the Middle to Upper Paleolithic boundary? Or did it build gradually, reaching modern levels not through a sudden advance but cumulative increments? The "it" in these questions can be replaced equally by image-making, technical innovation, and language—probably not by mere analogy, but as a real biological link. Is it conceivable, for instance, to imagine fully modern language without ability for image-making, or vice versa?

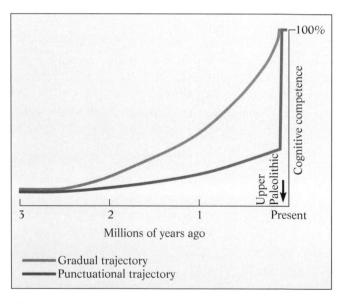

Two models dominate thinking on the origin of spoken language: one envisages a gradual rise in language capacity throughout hominid prehistory, while the second sees a punctuational increase in capacity with the evolution of modern humans.

The same may be asked of modern levels of innovation, and certainly of introspective consciousness.

How can the timing of modern language origins, however, be specifically addressed? First, through the products of language capability—albeit very indirectly, as language does not fossilize or impress itself directly on the archeological record except as writing, a very late development. And second, by means of evidence from the anatomical structures that produce language: the brain and the vocal tract.

A Recent, Abrupt Appearance?

For Randall White, evidence of various forms of human activity earlier than 100,000 years ago implies "a total absence of anything that modern humans would recognize as language." With the evolution of anatomically modern humans at this point, there were for the first time people who were neurologically equipped for language, if not yet socially or culturally equipped, argues White. Eventually the potential for language was fully realized, ushering in the Upper Paleolithic/Later Stone Age. "This revolution was effected by neurally capable populations of *Homo sapiens* in an exceedingly long struggle," White suggests. "By 35,000 years ago, these populations . . . had mastered language and culture as we presently know them."

In other words, White is proposing that the gap in language capabilities between modern *Homo sapiens* and their precursors was significant. There was no simple increment in language capabilities, the addition of a useful nuance or two. Instead, a major evolutionary shift occurred, endowing modern humans with a piece of mental machinery that demanded a long learning process in a heightened social and cultural context, to realize its full potential.

White lists seven areas of archeological evidence that, in his view, point to dramatic enhancement of language abilities coincident with the Upper Paleolithic, perhaps the result of a slow fuse lit some 65,000 years earlier. First is evidence of deliberate burial of the dead, which begins almost certainly in Neanderthal times, but becomes refined (with the inclusion of grave goods) only in the Upper Paleolithic. Second, artistic expression—image-making and bodily adornment—begins only with the Upper Paleolithic and Later Stone Age. Third, one sees a sudden acceleration in the pace of technological innovation and cultural change. Fourth, for the first time real regional differences in culture develop, an expression and product of social boundaries. Fifth, evidence of long-distance contacts—exotic objects traded between groups—becomes strong. Sixth, living sites significantly increase in size (complex language is a prerequisite for planning and coordination). Seventh, technology moves from the predominant use of stone to include other raw materials, such as bone, antler, and clay.

This combination of "firsts" in human activity looks impressive, although we have seen that some scholars question the punctuational appearance of certain of these elements. Nevertheless, for White and many others, including Richard Klein and Lewis Binford, the evidence is persuasive of a historically meaningful shift underlain by the appearance of complex, fully modern spoken language. Binford, for instance, sees no evidence of planning depth among Middle Paleolithic populations—little facility for predicting and organizing future events and activities, by contrast with Upper Paleolithic people. "If you ask, 'what makes this [change] possible,' you'd say, intelligence, yes," argues Binford. "But more important is language and, specifically, symboling, which makes abstraction possible. I don't see any medium through which such a rapid change could occur other than a fundamentally good, biologically based communication system." Klein, in essential agreement with this proposition, points to archeological evidence of a dramatic increase in hunting skills across the Middle to Later Stone Age transition in southern Africa.

It is, then, widely held among anthropologists today that language is a recent development in prehistory, coincident with the origin of modern humans. We will examine some of the archeological evidence that bears on the matter, beginning with technology.

Technology Points to Initial Gradual Evolution of Language

If the overall trajectory of technological change is examined from its beginnings 2.5 million years ago into the Upper Paleolithic, a significant pattern emerges, one that can be taken as support for the late origin of modern language. Glynn Isaac, who died in 1985, explored this pattern a decade and a half ago for a landmark conference on language origins held at the New York Academy of Sciences. "Asking an archeologist to discuss language is rather like a mole being asked to describe life in the treetops," he remarked; nevertheless, he pointed out that archeological evidence can reveal interesting information. For instance, in the long sweep of human technological history, the various elements of artifact assemblages became ever more numerous and ever more sharply defined, with major "improvements" occurring abruptly about 1.6 million years ago and some 250,000 years ago. These dates coincide with the appearance of, respectively, *Homo erectus* and archaic *sapiens*.

One interpretation of this pattern is that as human history progressed, stone-tool makers became more skilled, creating the same kinds of artifacts, but to ever more stringent standards. Choppers began to look more consistently like choppers, scrapers like scrapers, and so on. But, argued Isaac, "it is not necessarily true that the increase in complexity reflects an increase in the number of tasks performed with stone tools, nor are the fancy tools necessarily more efficient in an engineering sense. This is a point that has seldom been recognized."

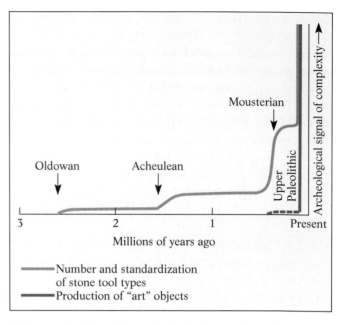

If the complexity of the archeological record may be taken as a guide, language capacity arose late in hominid prehistory.

If Isaac's argument is correct, what might have motivated the increasingly high standards of production? "My intuition in this matter is that we see in the stone tools the reflection of changes that were affecting culture as a whole," he suggested. "Probably more and more of all behavior, often but not always including tool-making behavior, involved complex rule systems. In the realm of communications, this presumably consisted more and more of elaborate syntax and extended vocabulary; in the realm of social relations, perhaps increasing numbers of defined categories, obligations, and prescriptions; in the realm of subsistence, increasing bodies of communicable know-how."

From Isaac's mole's eye view, therefore, there was an increase, however slow, in arbitrary ordering of society through time. It seems clear that, whatever they were doing, our *Homo* ancestors—*habilis* and *erectus*—had gone beyond what we see in the daily life of apes. This should not be too surprising, because over this period hominid brain size had increased to more than twice that of apes. Can this be taken as an archeological trace of steadily rising language capacity, beginning with the first species of *Homo* and continuing smoothly through to modern humans? No, said Isaac, because the trajectory of technological change shifts dramatically upward with the Upper Paleolithic, and this must be significant.

Between 2.5 million and 250,000 years ago, the pace of technological change (and, by inference, language-induced cultural change) closely parallels increase in brain size and therefore looks like biological evolution. After that, and particularly in the more recent Upper Paleolithic times, brain size remains the same while technological change gathers apace—a sure sign of the modern human mind (equipped with modern human language?) at work. In other words, real cultural evolution had kicked in. From the archeological evidence of stone-tool technologies, then, it is reasonable to deduce a gradual rise in some kind of language capacity, beginning with the origin of the genus *Homo*, but it is tempting to infer that the Upper Paleolithic era was ushered in by—was indeed the consequence of—a major enhancement in language abilities.

An Upper Paleolithic Signal

The same kind of pattern is to be seen in the second relevant area of archeological evidence, that labeled art. As we saw in the previous chapter, activities such as engraving, sculpting, and painting appear late in the prehistoric record, suddenly coincident with the emergence of innovation and rapid change that defines Upper Paleolithic tool technologies. Indeed, the

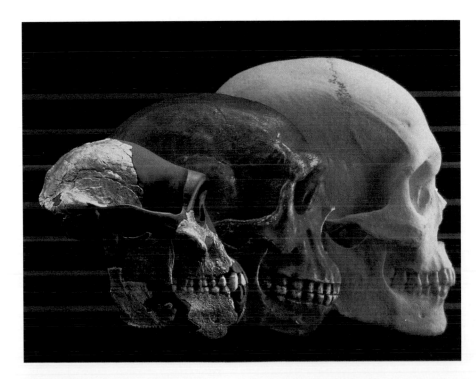

During human evolution the relative size of the brain increased threefold, as illustrated here by a comparison between Australopithecus afarensis *(left),* an early hominid, Homo erectus *(center), and* Homo sapiens.

prelude to Upper Paleolithic art is even less evident than is the technological precursor to Upper Paleolithic tool technologies. If the ability and penchant to produce images and decorative objects do relate to language abilities, therefore, the evidence of the archeological record would point to dramatic enhancement late in human prehistory.

Australian archeologist Iain Davidson has recently argued for this relation. "The making of images to resemble things can only have emerged prehistorically in communities with shared systems of meanings," he asserted in collaboration with his colleague William Noble, also at the University of New England. "Shared systems of meanings" are mediated, of course, through language. Davidson and Noble's main thrust, however, is that the development of language and the development of image-making are interdependent, each facilitating the other. Ergo, representational images in prehistory are the imprint of language in tangible form.

The technological and artistic realms of the archeological record, then, clearly signal a late, significant enhancement of spoken language abilities. This is the pattern supported by White, Binford, Klein, and others, and is consistent with other aspects of the record: of the seven archeological signals adduced by White above, technology and art surely represent that most arrestingly.

Ralph Holloway of Columbia University has worked extensively with hominid brain casts. Here he uses a craniometric stereoplotter to produce a detailed map of the surface of the brain cast.

Brain Anatomy Gives a Different Perspective

When we turn to other aspects of the prehistoric record, however, specifically to anatomical evidence of the brain and vocal tract, a different but less clear-cut conclusion emerges. Overall, this anatomical evidence can be read to imply a gradual emergence of language abilities throughout human evolution once brain expansion had begun with the appearance of the genus *Homo*, with no threshold-breaking shift late in prehistory.

As will become clear, extracting from the paleontological record anatomical evidence that is relevant to language abilities is no easy task. However, one measure that is relatively straightforward to make is the increase in the size of the brain. The australopithecine species, the earliest arrivals in the human family, had essentially ape-sized brains, around 450 ccs more or less. With the origin of the genus *Homo*, brain size began to increase. In *Homo habilis* (from something earlier than 2 million years ago to perhaps

1.7 million years ago), brain size was in the range of 600 ccs to 800 ccs. Its putative descendant, *Homo erectus* (from 1.7 million to 250,000 years ago), had a range of 850 ccs to 1100 ccs. The average for modern *Homo sapiens* is some 1350 ccs. (Neanderthals, lest it be forgotten, had a higher average than modern humans, about 1550 ccs.) What does this imply for the evolution of language?

Everything, according to Terrence Deacon of Harvard University. Based on studies of nerve connectivity in monkey brains, plus a reassessment of the changes in overall anatomy of the expanded human brain, Deacon concludes that language development effectively tracked the increase in brain size through human evolutionary history. "Moreover, language was the major cause, not just the consequence, of human brain evolution," he argues. "The neurological evidence indicates that language is not a recent addition to human culture but is on the order of two million years old." Deacon also suggests that, although their

speech would undoubtedly have been rather nasal and difficult for us to understand, Neanderthals were linguistically sophisticated and could have conversed with modern humans.

Ralph Holloway, a paleoneurologist at Columbia University, essentially agrees. "My bias is that the origins of human language behavior extend rather far back into the paleontological past and were nascent, but growing, during australopithecine times of roughly 2.5 to 3.5 million years ago," he says. "The form was undoubtedly primitive, but carried with it a limited set of sounds systematically used, and based on a well-known aspect of primate sociality, the ability, if indeed not penchant, for making noise." Holloway has spent many years studying the inner surfaces of fossilized human crania, mapping the faint signs of brain structure that become impressed naturally in the bone. He has long concluded that a humanlike gross anatomy of the brain—prominent frontal lobes, relatively small occipital lobes—is to be found in our

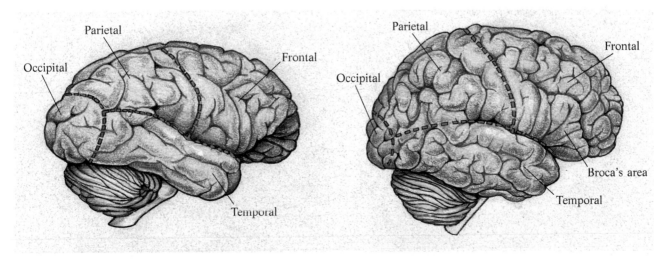

While the gross structure of chimpanzee (left) and human (right) brains is the same, differences are seen in the relative proportions of certain regions. In humans, the occipital lobe is relatively small whereas the parietal and frontal regions are large.

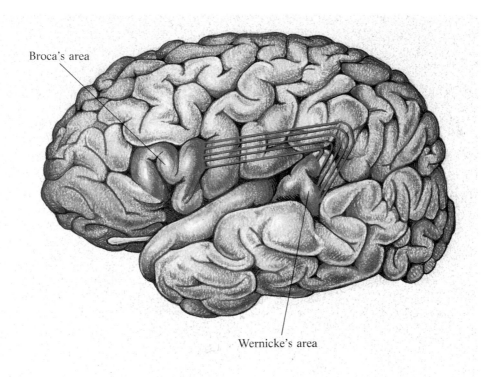

Classically, two brain areas have been associated with language function, namely Wernicke's area and Broca's area. In recent years it has become clear that many regions are involved in the production and comprehension of language, and these are distributed throughout the prefrontal region.

ancestors right from the beginning. The pattern contrasts with that of apes, in which the frontal lobes are smaller and the occipital lobes larger. This and other evidence from fossil brain patterns indicate, says Holloway, an early beginning to the development of spoken language in human prehistory.

Dean Falk, who, like Holloway, has made a study of human fossil brains, comes to the same overall conclusion. Although she does not agree with Holloway that the humanlike brain structure is present in australopithecines, she does concur that it is to be seen in earliest *Homo*. Falk, who works at the State

University of New York, Albany, recently remarked: "What it is going to take to *settle* the debate about when language originated in hominids is a time machine"; nevertheless, she continued, "if hominids weren't using and refining language I would like to know what they *were* doing with their autocatalytically increasing brains."

Those who have examined human fossils for evidence of language capabilities have traditionally looked for the presence of a small lump on the left (usually) side of the brain, near the temple: Broca's area, one of two major neural landmarks of language.

Such evidence is to be seen in skull 1470, a 1.8-million-year-old *Homo habilis* cranium from northern Kenya, as Holloway noted more than two decades ago. Recent investigations by several laboratories, however, including that of Marcus Raichle at Washington University Medical Center, have suggested that Broca's area is a less secure indicator of language than has been thought.

Nevertheless, evidence from brains—living and dead—has generally been interpreted to indicate gradual increase in language competence throughout human prehistory (at least, post-*Homo*) and that a giant cognitive leap did not occur with the appearance of modern humans. The argument, as we have

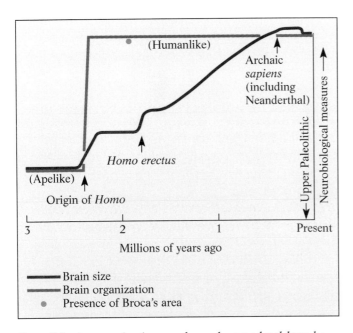

Overall brain organization reaches a human level long before brain size achieves the modern level, thus giving potentially conflicting evidence on emerging language capacity. Broca's area, which is regarded as less of a sure indicator of language capacity than it once was, appears with the origin of the genus Homo.

seen, has rested on brain size and overall organization. In this connection, Holloway has recently reevaluated the Neanderthal brain, once described as "deficient" and "primitive" compared with that of modern humans. In a paper titled "Poor Brain of Neanderthal: See What You Please . . . ," Holloway concluded that the brain was "fully *Homo*, with no essential difference in its organization compared to our own." The only difference is that it was bigger than ours. "Neanderthals did have language," he states firmly.

Message from the Voice Box

The brain is, of course, only part of the anatomical equipment responsible for the production of language. The language centers in the brain ultimately control the function of the various components of the vocal tract—the larynx, pharynx, vocal cords, tongue, lips, and jaws. Since much of this architecture is composed of soft tissue (muscle, cartilage, and skin), it vanishes from the fossil record—a pity, because the human vocal tract is highly distinctive, reflecting the very special job it performs. All is not lost, however, as Philip Lieberman, Edward Crelin, and Jeffrey Laitman have been discovering in recent years.

In all mammals apart from humans, the larynx is high in the neck, a position with two consequences. First, it allows the larynx to lock into the nasopharynx—the air space near the "back door" of the nasal cavity—an arrangement of some importance: it permits the animal to breathe and drink at the same time. Second, because the pharyngeal cavity—the sound box—is necessarily small as a result of the high larynx, the range of sounds the animal can make is quite limited. For typical mammals, vocalization therefore depends principally on the shape of the oral cavity and lips, which modify the sounds produced in the larynx.

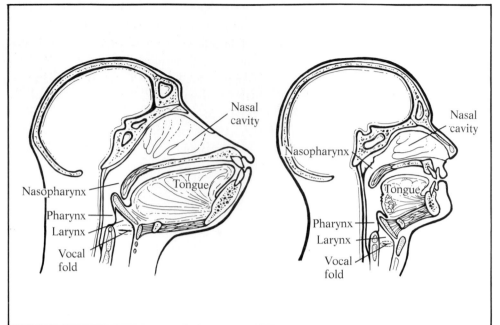

Nasal cavity

Nasopharynx

Tongue

Pharynx

Larynx

Vocal fold

Nasal cavity

Nasopharynx

Tongue

Pharynx

Larynx

Vocal fold

In the vocal tract of the chimpanzee (left), as in all mammals, the larynx is high in the neck, enabling simultaneous breathing and swallowing but limiting the range of sounds that can be produced in the pharyngeal space. The larynx in the human vocal tract (right) is lower, enabling a wide range of sound production but prohibiting simultaneous breathing and swallowing. The australopithecine vocal tract was like the chimpanzee's.

In humans, the larynx is much lower in the neck, with the consequence that humans cannot drink and breathe simultaneously; they constantly risk choking on food or liquid as it is swallowed. With a hazard of this magnitude, some great benefit must be conferred by the low position of the larynx. The most obvious is a much larger pharyngeal space above the vocal cords, allowing a greater range of sound modification. "The expanded pharynx is the key to our ability to produce fully articulate speech," explains Laitman at the Mount Sinai School of Medicine, New York.

Laitman and his colleagues discovered that human infants, as they develop, essentially recapitulate this event in our evolutionary history. Babies are born with the larynx high in the neck: the typical mammalian arrangement, and one necessary for them to nurse successfully without choking. After about one and a half years, the larynx begins to migrate down the neck, eventually reaching the adult position by about 14 years; speech development corresponds to this movement.

What evidence does this afford to anthropologists, who deal with fossilized crania devoid of all parts of the vocal tract? "During our investigations, my colleagues and I noticed that the shape of the bottom of the skull, or basicranium, is related to the position of the larynx," explains Laitman. "This is not surprising, since the basicranium serves as the roof of the upper respiratory tract." In the basic mammalian pattern, the bottom of the cranium is essentially flat. In humans it is arched. Although it is a delicate area of the cranium, often broken during burial and fossilization, the basicranium does from time to time remain intact, yielding clues to the anatomical structure of ancient vocal tracts. A flat basicranium may be taken to indicate the absence of language, whereas degrees of arching may be clues to various degrees of language competence.

Laitman and his colleagues therefore examined as many fossilized hominid crania with complete or near-complete basicrania as they could find. "The pattern we see is very interesting," says Laitman. "First, in all the australopithecines I've examined, you see a typical apelike basicranium. This indicates to me that it would have been impossible for them to have produced some of the universal vowel sounds that characterize human speech patterns. Second, the earliest time in the fossil record that you find a fully flexed basicranium is about 300,000 to 400,000 years ago, in what people call archaic *Homo sapiens*."

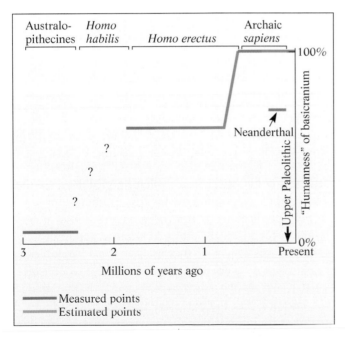

As an indicator of the "humanness" of the vocal tract, the basicranial shape implies the achievement of modern status with the origin of archaic sapiens (with an idiosyncratic reduction in Neanderthals).

When did basicranial flexion, the putative signal to an expanding vocal range, begin? "We can say it had already begun with *Homo erectus*," says Laitman. "We examined skull 3733, a lovely *Homo erectus* fossil from east Lake Turkana, about 1.6 million years old. The degree of flexion you see is about like that of a six-year-old human, so you could say that *Homo erectus* could produce a significant range of sounds, far greater than an ape can." The *erectus* people would, however, have been somewhat restricted in the vowel sounds they could produce, lacking, for instance, vowels as in boot, father, and feet. Nevertheless, if the inferences about the vocal tract are correct, it seems clear that early on in human evolution, language development was quite advanced. In some Neanderthals, incidentally, the basicranium is less arched than in *erectus*, implying a reversal of this evolution—an extraordinary event. Laitman suspects that the flattening of the Neanderthal basicranium might be related to the unusual cranial anatomy overall, particularly the "pulled-out" facial arrangement. The result might have been a rather nasal speech, perhaps lacking some modern vowels. From this evidence Lieberman, for instance, has argued that Neanderthals were less language competent than modern humans—a possible factor in their extinction, he has said. The question of how advanced spoken language was in Neanderthals, and whether they had actually regressed from the level achieved in *Homo erectus*, remains controversial.

What of earliest *Homo*—*Homo habilis*, the ancestor of *Homo erectus*? Had the larynx already begun its evolutionary descent, with language abilities of the *Homo habilis* beyond the merely nascent? Unfortunately, no *habilis* specimen has an intact basicranium, so for now at least, the question must remain moot.

The issue of a late, dramatic origin of modern human language as against a steady, incremental trajectory is therefore unresolved. Archeological evidence is generally interpreted to support the former,

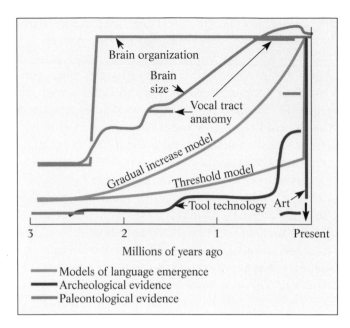

A comparison of all lines of evidence—paleontological and archeological—gives no consistent support for either the punctuational or gradual models for the emergence of language capacity.

paleontological evidence to support the latter. Clearly, the interpretation of evidence in one realm or the other is incorrect or incomplete.

Aping Language

In some respects, the dichotomy of interpretations in the archeological and paleontological realms parallels the theoretical dichotomy that exists over the underlying nature of human language. The discontinuity school, championed principally by the eminent linguist Noam Chomsky, sees language as a uniquely human trait with no direct evolutionary connection to ape brains; for the continuity school, however, human language is part of a cognitive continuum, rooted in our apelike ancestors and ultimately in the basic com-

municative and cognitive skills of our closest genetic relatives, the great apes.

For more than three decades, psychologists and other researchers have been addressing this theoretical divide, essentially by looking for the cognitive underpinnings of language in living apes, particularly chimpanzees. The research has been dogged by disputes over methodology, some of which was frankly sloppy, so that results could not be securely interpreted. Many researchers claimed to have demonstrated some language abilities in apes, but their interpretations could legitimately be challenged. Recently, however, one project on a pygmy chimpanzee (*Pan paniscus*), or bonobo, seems to have come to terms with methodological issues and produced convincing evidence for some neurological substrate to language.

Sue Savage-Rumbaugh, working at the Language Research Center of Georgia State University, has shown that Kanzi, a 10-year-old male bonobo, can comprehend complex spoken sentences—a first in ape-language studies. Kanzi picked up his now-extensive vocabulary of comprehension much as human infants learn language, by exposure to everyday discourse—in Kanzi's case, words spoken and indicated on a lexigram. (Previous ape-language projects typically involved rote learning of a few symbols at a time.) Kanzi has also shown some facility for structuring his utterances, which he produces via the 260-symbol lexigram. This facility, although limited to two- or three-word sentences, shows a logical consistency: for instance, the indication of activity followed by the object of that activity ("eat apple" or "tickle Kanzi"). For this reason it is best to consider what Kanzi displays as a protogrammar rather than to claim for him the extensive grammatical abilities required for propositional language. "It is reasonable to conclude that in Kanzi we are seeing the cognitive substrate to human language," says Savage-Rumbaugh, "limited, as you'd expect in a brain one third the size of a human brain, but definitely present." Although the trajectory of the later stages of language development are not directly addressed by this kind of research, her conclusion, if correct, would encourage the notion that such development might well have begun early in human history.

Why Did Language Evolve?

We now turn from the timing of the evolution of fully modern language to its function, the critical category for natural selection. Communication is the most obvious answer. Human spoken language is unprecedented among animals both for information content and rate of transmission. In our modern world, where information is all-pervasive, we constantly use language to communicate, so the communication hypothesis for language origins is seductive. And it is easy to imagine how useful language would have been to hunter-gatherers for organization and planning.

Nevertheless, to separate present utility from initial selection is a constant and challenging requirement in evolutionary theorizing. In recent years many scholars have come to the conclusion that however important a part language plays in human communication, language origins may have been linked to something quite different.

"The role of language in communication first evolved as a side effect in the construction of reality," proposes Harry Jerison, a neurologist at the University of California, Los Angeles, who has made a special study of brain evolution. "We can think of language as being merely an expression of another neural contribution to the construction of mental imagery." Brains throughout evolutionary history have been shaped to construct an inner world appropriate to a species' daily life. In amphibians, vision provides the principal element of that world; for reptiles, an acute sense of smell. For the earliest mammals, hearing was additionally important; and in primates, a mélange of sensory input creates a complete mental model of external reality. Humans, says

Jerison, have added a further component: language, or more precisely, reflective thought and imagery. Thus equipped, the human mind creates an internal model of the world that is uniquely capable of representing and coping with complex practical and social challenges. Inner reflection, not outer communication, was the facility upon which natural selection worked, argues Jerison. Language was its medium—and, at the same time, an efficient tool for communication. This hypothesis now has wide support.

The inner world of propositional language is intimately related to imagery (and, as we will soon consider, introspective consciousness). Imagery is a powerful analytical as well as creative medium. As Jerison says, "In hearing or reading another's words we literally share another's consciousness." If the creation of an inner world was the driving force of the evolution of reflective language, and if this was a major part of human cognitive development over millennia, perhaps the dramatic brain-size increase that occurred throughout human evolutionary history is explicable principally in terms of the evolution of language— just as Terrence Deacon argued from basic neurological evidence.

The principal alternative hypothesis explaining brain-size increase, that of ever-greater technological powers, seems to Jerison "to be an inadequate explanation, not least because tool making can be accomplished with very little brain tissue." By contrast, he says, "The production of simple, useful speech requires a substantial amount of brain tissue."

Language and introspective consciousness are inextricably linked, certainly in the way most of us experience them. Were they also linked in an evolutionary context, bound to each other as an object of natural selection? Human consciousness has been the target of philosophical discourse for centuries, often elevated to the level of mystical phenomenon. But in recent years many anthropologists and primatologists have come to view a facility for consciousness as a useful tool in the world of social primates from which humans evolved.

A pe Language

A little over a decade ago, ape-language studies appeared to have been discredited. Herbert Terrace, a psychologist at Columbia University, New York, stated in a landmark paper in *Science* (March 1980) that apes were doing little more than mimicking their instructors—a skill that, in different ways, rats and pigeons perform with ease. The apes displayed no sense of grammatical structure in their use of symbols, he said. The issue related to an intellectual dichotomy: on one hand, a continuity school viewed human language as part of a cognitive continuum from our apelike ancestors; on the other hand, a discontinuity school believed that language was a uniquely human trait with no direct evolutionary connection to ape brains.

Ape-language studies, begun in earnest in the 1960s, were anchored in the first of these two schools. Seeking to demonstrate in apes the existence of such elements of human language as the use of various kinds of symbols within a grammatical structure, researchers hoped to reveal the "cognitive substrate" of human language in our close relatives. To those in the discontinuity school, the most prominent figure of which was Noam Chomsky, ape-language researchers were searching for something that did not exist. If language is a uniquely human trait, no semblance of it could be discovered in apes, they argued. Claims to demonstrate continuity must therefore be illusory.

Sue Savage-Rumbaugh has been involved with ape-language studies since the early 1970s, first at the University of Oklahoma, where Roger Fouts had been training a chimp called Washoe, and later at Georgia State University and Yerkes Regional Primate Research Center. With concerns over the validity of ape-language work similar to those of Herbert Terrace, Savage-Rumbaugh moved to Yerkes, where she eventually developed an important new approach. While Terrace was looking for production of grammatical structure under strict laboratory conditions, Savage-Rumbaugh began to shift emphasis, both in technique and goals. Instead of putting the chimps through rote learning, building up the vocabulary a symbol at a time, she decided to take a more naturalistic approach. A large vocabulary of symbols would be employed from the beginning, used as language as speech is employed around human infants.

Some ape-language programs have employed American Sign Language, in which the animals are taught by repeatedly molding their hands and fingers to the correct shape. It is a laborious process and, believes Savage-Rumbaugh, one that interferes with communication. Instead, the Language Research Center system employs an extensive lexigram, a matrix of 256 geometric designs on a board. Instructors touch the symbols, which represent verbs, adjectives, and nouns, to create simple requests or commands. At the same time, the sentence is spoken, with the objective of testing aural comprehension of English.

Several attempts have been made to teach chimpanzees to use signs symbolically. Here we see a common chimpanzee with the lexigram system developed by Duane Rumbaugh and Sue Savage-Rumbaugh at the Language Research Center, Georgia State University.

By the age of two years Kanzi, a male pygmy chimpanzee, had picked up some half-dozen symbols and spoken words "just naturally as human children do," says Savage-Rumbaugh. Language acquisition in humans is relatively slow and is spread out over some half-dozen years. Over about the same period, Kanzi's language abilities changed too, becoming more sophisticated in both comprehension and structure. This similar, albeit slower and more limited, developmental process in pygmy chimpanzees is indicative of the presence of a cognitive substrate of language in apes, suggests Savage-Rumbaugh.

By the age of 10 years Kanzi had a vocabulary of some 200 words, expressed to him either as symbols or as speech. It was not the size of his vocabulary that was significant, however, but what the words apparently meant to him. Experimental psychologists are familiar with the prodigious feats of association that even the most humble animal may perform. Chimpanzees, being very smart, are likely to be able to behave in complex ways that might mimic language abilities but actually be mere association, the linking of sounds and symbols with objects in the absence of true understanding. Savage-Rumbaugh believes that Kanzi's abilities go beyond this and are in the territory of rudiments of language.

Recently, the Language Research Center team conducted a series of tests on Kanzi's comprehension, giving him sentences—requests to do things—delivered by someone out of his sight. Team members in the room with Kanzi wore earphones; they could not hear the instructions and therefore could not cue Kanzi, even unconsciously. None of the sentences was practiced, and each was different. The first were relatively simple: "Can you put the raisins in the bowl?" "Can you give the cereal to Karen?" Kanzi did them easily.

Savage-Rumbaugh and her colleagues, going a step further, discovered an interesting discrimination that Kanzi made as to the structure of certain instructions. Because the command "Go to the colony room and get the orange" might be thought simply to link colony room and the orange that the bonobo finds there, they added a complication. With an orange in front of Kanzi, the instruction was repeated. About 90 percent of the time Kanzi seemed uncertain, fumbled with the orange in front of him, then went to the colony room and fetched the orange from there. But if the instruction were phrased differently—"Get the orange that's in the colony room"—Kanzi had no hesitation. The phrase "that's in the" is key here. "This suggests to me that the syntactically more complex phrase is producing better comprehension than the simple one," says Savage-Rumbaugh.

Kanzi's word production through the lexigram system also increased with age, indicating some kind of developmental process. In collaboration with Patricia Marks Greenfield, a psychologist at the University of California, Los Angeles, Savage-Rumbaugh showed that in producing word combinations Kanzi not only was able to learn simple grammatical rules, but also invent his own rules. The rule he picked up from his caregivers was that in two-word utterances, action precedes object. Although during the first month of the study Kanzi employed no specific order for action and object symbols, during the last four months he did so to a statistically significant degree. "This developmental trend from random ordering to an ordering preference was also found for human children at the two-word stage," note the researchers.

These and other observations persuade Marks Greenfield and Savage-Rumbaugh that they are witnessing real evidence of the cognitive substrate that underlies human language. "The capacity for grammatical rules (including arbitrary ones) in Kanzi's semiotic productions shows grammar as an area of evolutionary continuity," they say. "We might prefer to speak of protogrammar rather than grammar. However . . . the comparative data are such that if we speak of bonobo rules as protogrammar, we should apply the same term to the 2-year-old child."

Language as a Social Tool

Just as language may be seen to have evolved as a tool for creating better models in the mind (Jerison's argument), so can consciousness. Specifically, consciousness may have evolved as an aid to understanding—and predicting more accurately—a complex social environment. Most organisms in the world face daily challenges of a fairly limited and repetitive kind: finding food, avoiding predators, locating mates. But for primates, particularly higher primates, life is significantly more complicated. Their highly developed sociality introduces elements of unpredictability unmatched in most organisms' lives. The forming and breaking of alliances with other individuals, the manipulation of others' behavior, the monitoring of others' behavior to avoid being manipulated—this intense social nexus in the life of higher primates puts extreme demands on intellectual abilities. "It asks for a level of intelligence which is unparalleled in any other sphere of living," suggests psychologist Nicholas Humphrey, a leading investigator of the nature of consciousness. "Like chess, a social interaction is typically a transaction between social partners. One animal may, for instance, wish by his own behavior to change the behavior of another; but since the second animal is himself reactive and intelligent the interaction soon becomes a two-way argument where each 'player' must be ready to change his tactics—and maybe his goals—as the game proceeds. Thus, over and above cognitive skills which are required merely to perceive the current state of play, the social gamesman, like the chess-player, must be capable of a special sort of forward planning."

Consciousness therefore began to evolve in this highly competitive social setting, Humphrey argues. Working like an "inner eye," introspective consciousness allows an individual to guess how another might behave in a certain situation through self-aware experience of that same situation. "Consciousness provides me with an explanatory model, a way of making sense of my behavior in terms which I could not devise by any other means," explains Humphrey. "The introspectionist's privileged picture of the inner reasons for his own behavior is one which he will immediately and naturally project on other people. He can and will use his experience to get inside other people's skins"—and play more effective social chess.

Consciousness is particularly valuable in primate societies for predicting the behavior of other individuals in order to maximize reproductive opportunities. Once social skills become significant in a species, selection will further sharpen them. "In these circumstances there can be no going back," says Humphrey. "An evolutionary 'ratchet' has been set up . . . to increase the general intellectual standing of the species."

By this argument, higher primates probably have a degree of self-awareness; and there is some experimental evidence that they do. A level of consciousness almost certainly operates in the minds of apes, perhaps aided by what Savage-Rumbaugh describes as "the underlying substrate for language," mental machinery that seems to promote organization of thought processes. But the emergence of spoken language in hominids—specifically in *Homo*—can be seen as part of the process of enhancing consciousness to the exquisite clarity of human introspection.

If indeed language and consciousness evolved in the context of enhancing individuals' ability to operate within the complexities of primate social life, it is clear that the consequences go far beyond that initial purchase of natural selection. Disparities are often seen between present function and initial utility in evolutionary systems. One of the most extraordinary consequences of the evolution of introspective consciousness, moreover, may be the emergence of mythology.

The Fount of Myths

Mythology—more or less elaborate stories of how the world came to be and how it operates—is a mark of the modern human mind, the creation of worlds shared in the medium of language. Central to all (and the core of the world's religions) is a creation mythology—an explanation of how a people came to be. "Man, apparently, cannot maintain himself in the universe without belief in some arrangement of the general inheritance myth," observed Joseph Campbell. "Every people has received its own seal and sign of supernatural designation, communicated to its heroes and daily proved in the lives and experiences of its folk." Harvard biologist Edward Wilson makes the same point, with a more analytical slant: "The predisposition to religious belief is the most complex and powerful force in the human mind and in all probability an ineradicable part of human behavior. . . . It is one of the universals of social behavior, taking recognizable form in every society from hunter-gatherer bands to socialist republics."

How might this force spring from the font of consciousness? "When humans became aware of themselves as individuals, as having feelings and motivations, they not only attributed similar feelings to other humans but to other animals and to inanimate objects in the world," suggests psychologist Gordon Gallup, at the State University of New York, Albany. "Just as the behavior of other individuals might be understood and sometimes manipulated, there was born the notion that the rest of the world might also be understood, sometimes manipulated. Often, though, the world seemed full of enigmatic forces, mysteries that carried important powers."

This line of argument, briefly presented here, leads one to see how, from the moment consciousness burned brightly in the minds of modern humans, there has been a universal urge to account for the rest of the world—to tell stories of how things came to be, which forces were good, which evil, how they might be influenced. "Consciousness," observed Humphrey, "which developed initially to cope with local problems of interpersonal relationships, has in time found expression in the institutional creations of the 'savage mind'—the high rational structures of kinship, totemism, myth and religion which characterize primitive societies."

With awareness of self comes, inevitably, awareness of death—the Ultimate Concern, as Theodosius Dobzhanzky spoke of it. Death awareness, and the practice of burial that sometimes goes along with it, provide the archeologist with the possibility of gleaning from the past something of the level of consciousness in our ancestors' minds. The discovery of the fossilized remains of an individual obviously committed to a grave, perhaps accompanied by offerings of various kinds, provides eloquent testimony to a highly developed sense of self and death. Evidence of such moments, eagerly sought—and found—by archeologists over many decades, has led to the conclusion that the Neanderthals had a fully developed human consciousness—recall the opening scenario of our Prologue. Although virtually no evidence of ritual burial is to be found earlier than Neanderthal times, the practice has been widely accepted as advanced among these people. Recently this viewpoint has been challenged, however; the evidence of ritual burial has been called overinterpreted.

Evidence from the Grave

It is worth remembering how very Western is the practice of burial. Many cultures pursue elaborate death rituals that ensure a safe passage of the departing spirit, yet leave no archeological trace. Cremation on a funeral pyre is one obvious example, as is leaving the corpse ceremonially on a mountainside to be devoured by scavengers, or suspended on a platform where decomposition precludes fossilization. Absence of evidence of burial therefore does not necessarily preclude death awareness, although archeologists are on safer ground with positive rather than negative evidence.

There is no question that Upper Paleolithic people buried their dead, often with decorated objects and other artifacts. Such burials are not common, however, and in the Magdalenian era—the height of Ice Age art—none is known. The real question is, when did such practices begin? And when did death awareness first arise in the human mind? There are several dozen putative Neanderthal burials, almost always in caves; by contrast with Upper Paleolithic burials, however, grave goods are virtually absent.

In this 20,000-year-old burial at Sungir, 130 miles east of Moscow, strands and beads and a headband of mammoth ivory and fox teeth can be seen.

Nevertheless, does not the practice of burial alone—consider the old man of Shanidar, accompanied, as we described in the opening pages of this volume, with garlands of flowers—surely bespeak a human level of consciousness?

Not if the evidence is viewed dispassionately, argues Robert Gargett, an anthropologist at the University of California, Berkeley. In the fossil record prior to Neanderthal times, there is a virtual absence of complete skeletons: effectively, two in a 3-million-year sweep of human history, including the famous "Lucy" skeleton, a member of the earliest known hominid species, *Australopithecus afarensis*. After the Neanderthals evolved, complete skeletons become a feature of the record, if not exactly common; more than 20 are known. Some scholars cite this fact alone in favor of deliberate burial by Neanderthals, since burial greatly increases the chance that a skeleton will remain intact during the fossilization process. Gargett suggests that the proliferation of whole skeletons from Neanderthal times is merely the result of the remains having less time to be disrupted by natural, geological, or other disturbances. His main argument against Neanderthal burial, however, focuses on the putative cases themselves.

"If you look carefully at the evidence cited in favor of burial, such as shallow graves, arrangement of animal horns with a skeleton and so on, you can always come up with an explanation that involves only natural processes," he says. For instance, water gently swirling around a body or skeleton in a ground surface cave may excavate a shallow depression mistaken for a deliberate grave. "In many cases simple and likely explanations have been ignored in favor of complex scenarios invoking enigmatic purposeful behavior."

At Shanidar, the notion that the old man had been laid to rest on a bed of flowers "ignores the most probable agent for the deposition of flowers: wind," says Gargett. "Because it was already believed that purposeful disposal was a possibility, the discovery of flower pollen convinced the investigators that

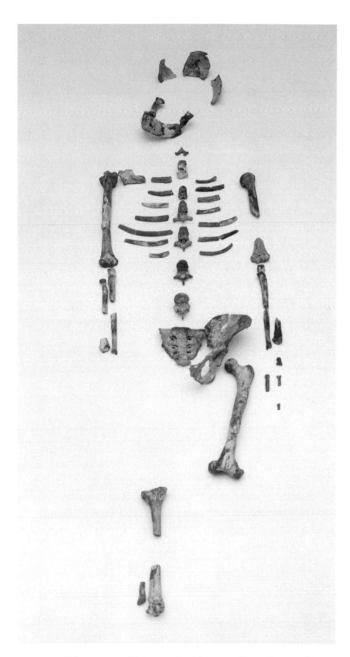

This 3-million-year-old Australopithecus afarensis *specimen named Lucy was found in Ethiopia in 1974 and represents one of only two partially complete skeletons of hominids earlier than Neanderthal. The second specimen is that of the Turkana boy, a 1.6-million-year-old* Homo erectus *from Kenya.*

Neanderthals had burried their dead with flowers." Such "self-delusion" has been common in excavations where intentional burial is suspected, claims Gargett. In a sharp riposte, Arlette Leroi-Gourhan says: "Gargett imagines the wind's having blown the flowers just into the Neanderthal burial soil and having chosen bright-colored flowers belonging to five different genera. It is a pity that he has constructed his argument without considering the dispersion of pollens and without reading my paper on the subject."

In his own paper, published in 1989 in *Current Anthropology*, Gargett concluded that "the working hypothesis should be that Neanderthals did not bury the dead or otherwise transform sediments in the course of performing ritual." A number of scholars responded in the same issue of the journal with comments ranging from "a very well-founded and sound reinterpretation" to "we have difficulty finding scientific merit in this paper." Erik Trinkaus, an eminent Neanderthal scholar at the University of New Mexico, agrees that Gargett is correct to say that evidence adduced in support of ritual burial has been "overinterpreted in some cases," but suggests that Gargett goes too far in discounting all such claims. "In other cases," Trinkhaus notes, "no matter how critical one is, intentional burial is the only reasonable conclusion. Neanderthals did bury their dead. What you make of that is another matter."

An Unending Search

If Trinkaus is correct in concluding that ritual burial was indeed part of Neanderthal culture—and most anthropologists would accede to that interpretation of the evidence—then it seems safe to assume that awareness of death was an experience in the Neanderthal mind. But how well-developed an experience was it? The attempt to explore another's subjective experience is in a human sense frustrating, particularly when that experience is many tens of millennia

An Ancient Ritual That Wasn't

The practice of burial, particularly with attendant ritual, is correctly taken as indicative of human spirituality. Archeological evidence of such activity has therefore been sought with the aim of pinpointing when the truly human mind arose. For the archeologist, burial has the merit of high potential visibility in the prehistoric record.

Prior to the appearance of Neanderthals, there is no evidence of burial; with the origin of modern humans, burial practices can be clearly identified. Since the turn of the century at least 30 of the 200 or so Neanderthal discoveries made have been said to reveal mortuary practice, including such activities as positioning the corpse, adding artifacts and other animal bones to the grave, and arranging stones significantly.

Robert Gargett's reexamination of the reports of six famous instances of putative burial—La Chapelle-aux-Saints, Le Moustier, La Ferrassie, Teshik-Tash, Regourdou, and Shanidar—focused, as in virtually all cases of putative burial, on Neanderthal bones recovered from caves or rock shelters. Gargett points out that the processes of deposition, erosion, and redeposition are often dynamic in caves, making interpretation of evidence extremely problematic. For instance, roof falls may give the impression that a corpse has been covered by a stone slab or that bones have been broken in a deliberate, ritualistic way, and the use of human-occupied caves by carnivores, particularly hyenas, also serves to complicate the evidence.

Although many scholars agree with Gargett that the onus of proof should be on those who claim to see evidence of burial, the majority consider that he has gone too far in denying compelling evidence in favor of the custom. One general fact that argues for Neanderthal burial is the extraordinary number of partial or complete skeletons known. And yet one of the most famous putative examples of Neanderthal death ritual was recently scrutinized and failed the test.

In 1939 a Neanderthal skull was discovered in a limestone cave on the southern slopes of Monte Circeo, some 100 kilometers south of Rome. The skull was said to show signs of ritual mutilation: the foramen magnum had been enlarged for access to the brain, and the skull itself was reported to have been resting on its crown, base upwards, in the center of a ring of stones. Damage to the right side of the skull was the result of ritual murder, it was argued. For more than half a century pictorial reconstructions of the supposed ritual cannibalism have appeared in popular anthropology volumes as evidence of Neanderthal spirituality. In 1991 Mary Stiner of the University of New Mexico studied the fossil assemblage associated with the skull; Tim White of the University of California, Berkeley, and Nicholas Toth of Indiana University added an assessment of the skull itself. Their conclusions were very different from earlier interpretations.

Stiner tallied the animal species whose bone fragments lay at the same level as the Neanderthal skull, and she looked for evidence of carnivore activity, such as gnaw marks in the surface of the bone. She did the same for levels below the skull, in which Mousterian tools had been found. These data she compared with bone assemblages from a nearby known Neanderthal site and from a hyena den. Although the lower, tool-bearing levels matched the pattern for a human occupation, which showed a low accumulation of animal bones and little gnawing, the assemblage at the skull level was indistinguishable from that of the hyena den, which displayed an abundance of bones, many with gnaw marks and punctures from canine

removed. Even more daunting is the interpretation of a prehistoric record that is at best incomplete and at worst ambiguous.

As we have seen, the tangible evidence may be interpreted in very different ways. Did modern reflective language arise recently and dramatically, or in a long, incremental process? Did the people of Shanidar bury their shaman in the way our opening paragraphs suggested, or does that scenario merely impose modern human notions upon highly equivocal

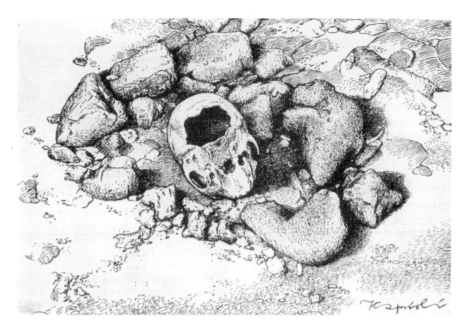

Alberto Blanc's reconstruction of the Neanderthal cranium within a circle of stones in the Monte Circeo Cave, Italy, which, he argued, indicated ritual cannibalism. Recent analysis has thrown doubt on the validity of this interpretation.

on its left side, not directly on its crown as originally reported. Marks left by implements were those of drills and picks employed during the excavation and preparation of the fossil—not 60,000 years ago, when the individual died. There were no ancient cut marks, no polish, no flaking—in fact, no evidence that the cranium had been modified by human hands until it was recovered as a fossil. The damage to the left side of the skull, which had been interpreted as evidence of ritual murder, was consistent with carnivore chewing, suggested White and Toth.

All three researchers pointed out that the evidence for a stone circle, tenuous at best, is more properly described as an irregular scatter of stones such as can be found anywhere on the cave floor.

The case of Monte Circeo does not disprove the notion that Neanderthals were capable of mortuary ritual, merely that it was not apparent in this instance. But the rigorous reexamination of a case that had worked its way deep into anthropological lore has been salutary. It gives an insight into the kind of evidence that is relevant to a reliable interpretation of past human activity. It is also a reminder that the urge to see evidence of the human spirit in our recent ancestors is strong—sometimes too strong.

teeth. She concluded that the Neanderthal skull probably arrived in the cave by the same means as other bones there: in the jaws of a hyena.

For their study, White and Toth familiarized themselves with the anatomy of ritual cannibalism by examining crania from the American Southwest and from Melanesia. The defleshing of skulls and their preparation for ritual practice leaves many characteristic signs, including marks in the surface bone left by stone knives; polish; and flaking bone on the interior of the skull: the result of exterior blows, particularly around the foramen magnum.

After examining the Neanderthal cranium the researchers noted first that the distribution of stain and a partial deposit of cave coral (calcium carbonate) indicated that it had lain

evidence? These remain the greatest archeological questions.

The origins of language—the human medium of thought, of communication, and of the many elements of culture and consciousness—and its role in the origin of modern humans remain an enigma. In the absence of Dean Falk's time machine, it is possible that we will never fully know the answers to these questions. It is equally certain that we will always want to know.

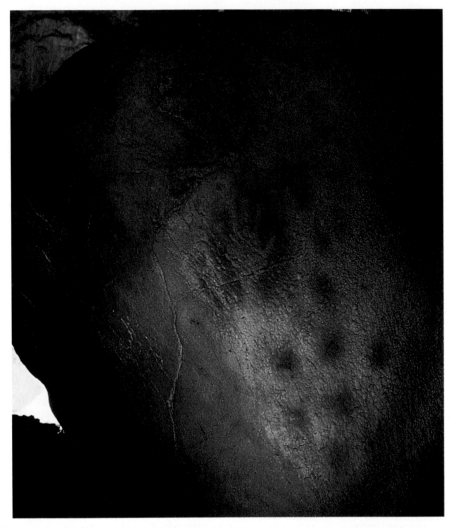

Negative hand stencils, produced by blowing pigment around the edge of the hand as it is held against a rock surface, are common in all prehistoric art. This 18,000-year-old example is from the cave of Peche Merle, in the Lot region of France.

GLOSSARY

Absolute dating: a date for a fossil or artifact derived by measuring the age of the sediments in which they are found, usually by radiometric methods.

Adaptation: the evolutionary process by which an organism becomes fitted to its environment; a character that has evolved under natural selection for some particular activity.

Acheulian: name applied to a type of stone-tool industry characterized by large bifaces including handaxes; began about 1.5 million years ago and continued in Africa and Eurasia until about 200,000 years ago.

Allele: alternative form of a gene, for example different eye colors; all genetic loci comprise two alleles, whose affects may differ depending on whether there are two identical alleles or two different alleles. *See* Polymorphism.

Anthropoid: the informal name for the suborder of primates that includes monkeys, apes, and humans.

Analogy: similarities among organisms based on convergent evolution (contrast with homology).

Anthropology: the study of humankind, including human evolution, human variability, and human behavior, past and present.

Apes: the groups of primates that includes chimpanzees, gorillas, and the orangutan (the great apes); and the gibbons and siamangs (the lesser apes).

Arboreal: tree-living.

Archaic Homo sapiens: a term applied to hominids living between from 400,000 years ago until the appearance of modern humans; these hominids may belong to one species or to several, a matter over which there is considerable disagreement.

Archeology: study of prehistoric human cultures.

Artifact: object made by humans.

Assemblage: one of several terms used to describe the set of artifacts made by a hominid species; other terms include industry, culture, and tradition.

Aurignacian: the first major culture of the Upper Paleolithic period, in Europe, named after the Aurignac rockshelter in the French Pyrenees.

Australopithecine: the common term for a member of the genus *Australopithecus*.

Basicranium: the underside of the cranium.

Biostratigraphy: dating method based on evolutionary changes within an evolving lineage.

Bipedalism: upright walking on the two hind legs, as humans' habitual mode of locomotion.

Blade: a flake whose dimensions are at least twice as long as it is wide.

Bottleneck (population): occurs when population numbers crash to a low level, from which recovery occurs.

Broca's area: the region in the human brain lying toward the lower back of the (usually) left frontal cortex; once considered as a major language center but now recognized as one of many brain regions involved in language; named after the French surgeon Paul Broca (1824–1880).

Calvarium: the cranium minus the mandible.

Carnivore: a meat-eating animal.

Cerebral cortex: the outer layer of the cerebral hemispheres.

Cervical: neck region of the spine.

Chatelperronian: the Upper Paleolithic tool culture that appeared at the transition between Mousterian and Aurignacian traditions, with characteristics of both; it may have been the product of cultural contact between archaic and modern humans; named after the cave site of Chatelperron in Allier, central France.

Clade: a group of species that contains the common ancestor of a group and all its descendants.

Cladistics: the school of evolutionary biology that seeks relationships among species based on the polarity (primitive or derived) of characters.

Cladogram: a diagrammatic representation of species relationships. *See* Cladistics.

Classification: the practice of assigning species to categories of relationship.

Cleaver: a large, bifacially flaked stone tool with a long cutting edge, like an axe; often found with handaxes of the Acheulian tradition.

Cranium: the complete skull (brain case, face and palate, and lower jaw). *See* Postcranium.

Culture: the sum total of human behavior, including technological, mythological, aesthetic, and institutional activities.

Cusps: the conical projections on the surface of teeth.

Cut mark: a usually elongated mark on a bone made by the application of an artifact.

Dental arcade: the shape of the tooth row.

Derived character: a character acquired by some members of an evolutionary group, and therefore serves to unite them in a taxonomic sense and distinguish them from other species in the group (contrast with primitive character).

Diastema: gap between the lateral incisor and canine, particularly in the upper jaw.

Dimorphism: two distinctive forms; for instance as in canine dimorphism or body-size dimorphism.

Diurnal: active during daylight hours.

Dorsal: pertaining to the back of an animal (contrast with ventral).

Ecosystem: the collection of species of plants and physical terrain that form a specific system, such as a savanna, a rain forest, a woodland.

Encephalization quotient: a measure of brain size relative to the size of the body.

Endocast: the impression of the inner surface of the brain case; can be natural or experimentally made.

Ethnographic analogy: an analogy between the behavior of living or recent societies and that of prehistoric societies.

Ethology: the study of the social behavior of animal species in the natural environment.

Fauna: the animal component of an ecosystem.

Faunal correlation: the dating of a site by the similarity of its fossil fauna to that of another site of known age.

Femur: thigh bone.

Fibula: one of the two bones in the lower leg. *See* Tibia.

Flora: the plant component of an ecosystem.

Folivore: a leaf-eating animal.

Foramen magnum: the opening at the base of the skull through which the spinal cord passes.

Frugivore: a fruit-eating animal.

Gene flow: transmission of genes between populations through intermating.

Gene pool: all the genes of a population at a given time.

Genetic drift: genetic changes in a population as a result of random effects rather than natural selection.

Genome: the full complement of a species' genes and associated DNA.

Genotype: the genetic makeup of an organism (contrast with phenotype).

Geochronology: term for dating methods in geology.

Gracile: small, lightly built. *See* Robust.

Gradualism: mode of evolution that involves steady accumulation of small changes (contrast with punctuationalism).

Handaxe: stone tool with a long axis and two cutting edges converging to a pointed or rounded end; typical of the Acheulian tradition.

Holocene: geological epoch that began with the end of the Pleistocene, 10,000 years ago.

Hominid: the informal name for the Hominidae, the human family, as currently classified.

Hominoid: humans and apes.

Homology: similarities of structures between species as a result of common ancestry.

Homoplasy: a character shared between species as a result of convergent evolution.

Humerus: upper arm bone.

Industry: one of several terms to describe the set of artifacts made by a hominid species; other terms include assemblage, culture, and tradition.

Innominate: the fused half-portion of the pelvis; contains three bones—the ilium, ischium, and pubis.

Knuckle walking: quadrupedal mode of locomotion, on the knuckles of the hands and the soles of the feet, by chimpanzees and gorillas.

Larynx: the organ in the throat that contains the vocal cords; the voice box.

Later Stone Age: cultural period beginning about 40,000 years ago and ending after 10,000 years ago, in sub-Saharan Africa.

Levallois technique: toolmaking technique, originated in Africa 200,000 years ago, in which the core is shaped from which flakes of predetermined size and shape can be removed; characteristic of Mousterian technologies.

Lineage: an evolutionary line linked by common ancestry.

Lower Paleolithic: human cultural period, beginning with the first appearance of stone tools and ending about 200,000 years ago, usually referring to Europe (later part) and North Africa; the equivalent in sub-Saharan Africa is Early Stone Age.

Lumbar: lower portion of the spine, from waist down.

Magdalenian: a major Upper Paleolithic culture of Europe, extending from about 17,000 to 12,000 years ago; named after the rock shelter of La Madeleine in the Dordogne region of France.

Mandible: lower jaw.

Manuport: stone on an archeological site that could not have occurred naturally there, and must have been transported by humans.

Maxilla: upper jaw.

Middle Paleolithic: cultural period, beginning about 200,000 years ago and ending 40,000 years ago, in North Africa and Europe; in sub-Saharan Africa the equivalent is Middle Stone Age.

Mitochondria: small organelles responsible for energy metabolism of the cell; contain their own, small genome.

Mitochondrial Eve (model): hypothesis that argues for a recent African origin of modern humans, followed by movement into the rest of the Old World and total replacement of existing archaic populations; based on genetic evidence. *See* Multiregional Evolution; Out of Africa.

Molars: cheek teeth, following the premolars.

Molecular clock (biological): the concept of using accumulated mutations within the genome as a measure of the passage of time.

Morphology: the physical form of an organism.

Mosaic evolution: term applied to the phenomenon of different rates of evolutionary change in different parts of the body.

Mousterian: name given to a European stone-tool industry characterized by flakes made from prepared cores; began 200,000 years ago and continued until 40,000 years ago.

Multiregional Evolution (model): hypothesis that argues for a more or less simultaneous emergence of modern humans from established populations throughout the Old World; based mainly on fossil evidence. *See* Mitochondrial Eve; Out of Africa.

Neolithic: the New Stone Age, usually associated with the beginnings of agriculture, 10,000 years ago.

Nucleotide: basic unit of DNA.

Occiput: the rear bone of the skull.

Oldowan: name given to a stone-tool industry characterized by flakes and chopping tools produced by hard-hammer percussion of small cobbles; began 2.5 million years ago and continued in parts of Africa and Asia until 200,000 years ago, where the industry is more properly called chopping tool assemblages.

Omnivore: a species that includes a range of food types in its diet.

Order: a category of classification; a subdivision of Class.

Out of Africa (model): hypothesis that modern humans originated first in Africa and then migrated into the rest of the Old World; based principally on fossil evidence. *See* Mitochondrial Eve; Multiregional Evolution.

Paleoanthropology: the multidisciplinary study of human evolution.

Paleolithic: the Old Stone Age, starting with the first appearance of stone tools, 2.5 million years ago, and ending with the origins of agriculture, 10,000 years ago. *See* Lower Paleolithic; Middle Paleolithic; Upper Paleolithic.

Paleomagnetism: dating method based on the periodic reversal in the direction of the Earth's magnetic poles.

Paleontology: study of fossils and the biology of extinct organisms.

Parsimony (technique): method for seeking the most likely set of relationships among species/populations as a way of establishing their history.

Phenotype: the physical characteristics of an organism. *See Genotype.*

Phyletic change: the evolution of a new species through the gradual change of an existing species, resulting in no increase in species diversity.

Phylogeny: evolutionary history of a group of related organisms; family tree.

Pharynx: throat above the larynx.

Pleistocene: the first epoch of the Quaternary, beginning about 2 million years ago and ending 10,000 years ago.

Pliocene: the final epoch of the Tertiary, from about 5 million to 2 million years ago.

Polymorphism (genetic): variant of a gene, which may or may not have a phenotypic effect. *See* Allele.

Potassium–argon dating: a radiometric technique for dating volcanic rock, based on the decay of potassium-40 to argon-40; a major dating tool in paleoanthropology.

Postcranium: all of that part of the skeleton that excludes the cranium.

Prehistory: that part of human history that took place prior to written records.

Primates: order of placental mammals that includes prosimians and anthropoids.

Primitive character: a character that was present in a common ancestor of a group and is therefore shared by all members of that group (contrast with derived character).

Prognathous: having jaws protruding in front of the line of the upper face.

Prosimian: common term for the suborder of primates that includes lemurs, lorises, and tarsiers.

Punctuationalism: mode of evolution in which changes are concentrated into brief periods (contrast with gradualism).

Quaternary: the second period of the Cenozoic; includes Pleistocene and Holocene.

Radiometric: methods of dating based on decay of radioactive isotopes, such as potassium–argon and carbon-14 dating.

Radius: one of the two bones in the lower arm. *See* Ulna.

Robust: large, heavily built. *See* Gracile.

Sagittal crest: ridge running along the top of the skull, as an enlargement of area for muscle attachment.

Sexual dimorphism: the state in which some aspect of a species' anatomy consistently differs in size or form between males and females.

Solutrean: a major Upper Paleolithic culture, extending from 23,000 to 18,000 years ago; named after the site of Solutré in eastern France.

Speciation: the evolution of new species through the splitting of an existing lineage, thus increasing species diversity.

Stone Age: the earliest period of human culture, from about 2.5 million years ago until the first use of metal, about 5000 years ago; divided into Old Stone Age and New Stone Age.

Stratigraphy: sequential layering of deposits.

Systematics: the science of classification, as taxonomy.

Taphonomy: the study of the processes that impinge on a carcass at death, which may include slow burial and eventual fossilization.

Taxonomy: classification of organisms according to evolutionary relationship.

Terrestrial: ground-living.

Thoracic: upper region of body, between the waist and neck.

Tibia: one of the two bones in the lower leg. *See* Fibula.

Tuff: solidified layer of ash from volcanic eruption.

Upper Paleolithic: the cultural period beginning about 40,000 years ago and ending 10,000 years ago, with reference to North Africa and Europe; the equivalent period in sub-Saharan Africa is the Later Stone Age.

Ulna: one of the two bones in the lower arm. *See* Radius.

Wernicke's area: the region of the human brain associated with comprehension of speech, located in the upper part of the temporal cortex; named after the German neurologist Carl Wernicke (1848–1905).

Zygomatic arch: the cheek bone.

FURTHER READINGS

There are several books in the general press that relate to human evolution. These include:

Johanson, Donald, and James Shreeve. *Lucy's Child*. William Morrow, 1989.

Leakey, Richard, and Roger Lewin. *Origins Reconsidered*. Doubleday, 1992.

Lewin, Roger. *Bones of Contention*. Simon & Schuster, 1987.

Falk, Dean. *Braindance*. Henry Holt, 1992.

Reader, John. *Missing Links*. Little Brown, 1981.

Schick, Kathy D., and Nicholas Toth. *Making Silent Stones Speak*. Simon & Schuster, 1993.

Trinkaus, Erik, and Pat Shipman. *The Neanderthals*. Alfred Knopf, 1993.

More technical volumes on modern human origins are numerous and include:

Bräuer, Gunther, and Fred Smith, eds. *Continuity or Replacement: Controversies in* Homo sapiens *Evolution*. Rotterdam: A. A. Balkema, 1992.

Mellars, Paul, and Chris Stringer, eds. *The Human Revolution*. Princeton University Press, 1989.

Trinkaus, Erik, ed. *The Emergence of Modern Humans*. Cambridge University Press, 1989.

Here is a more detailed list of additional readings:

Prologue:

Leroi-Gourhan, A. The flowers found with Shanidar IV, a Neanderthal burial in Iraq. *Science* 190 (1975):562–564.

Solecki, R. Shanidar IV, a Neanderthal flower burial in northern Iraq. *Science* 190 (1975):880–881.

Chapter 1: Prelude to Homo sapiens

Aiello, L. C. Pattern of stature and weight in human evolution. *American Journal of Physical Anthropology* 81 (1990):186–187.

Andrews, P., and L. Martin. Cladistic assessment of extant and fossil hominoids. *Journal of Human Evolution* 16 (1987):101–118.

Cartmill, M., et al. One hundred years of paleoanthropology, *American Scientist* 74 (1986):410–420.

Hill, A., and S. Ward. The origin of the Hominidae. *Yearbook of Physical Anthropology* 31 (1988): 49–83.

Lovejoy, C. O. Evolution of human walking. *Scientific American* (November 1988).

Pilbeam, D. Descent of the hominoids. *Scientific American* (February 1984): 84–96.

Rodman, P. S., and H. McHenry. Bioenergetics of hominid bipedalism. *American Journal of Physical Anthropology* 52 (1980):103–106.

Simons, E. Human origins. *Science* 245 (1989):1343–1350.

Tattersall, I. Species concepts and species recognition in human evolution. *Journal of Human Evolution* 22 (1992):341–349.

Wood, B. Origin and evolution of the genus *Homo*. *Nature* 355:783–792.

Chapter 2: Prelude to the Modern Debate

Aitken, M. J., et al., eds. The origin of modern humans and the impact of chronometric dating. *Philosophical Transactions: Biological Sciences* 337 (1992):125–250.

Brace, L. The fate of the classic Neanderthals. *Current Anthropology* 5 (1964):3–43.

Graves, P. New models and metaphors for the Neanderthal debate. *Current Anthropology* 32 (1991):255–274.

Grün, R., and C. B. Stringer. Electron spin resonance and the evolution of modern humans. *Archeometry* 33 (1991):153–199.

Hammond, M. A framework of plausibility for an anthropological forgery. *Anthropology* 3 (1979): 47–58.

Hammond, M. The expulsion of the Neanderthals from human ancestry. *Social Studies in Science* 12 (1982):1–36

Reader, J. *Missing Links*. Little Brown, 1981.

Spencer, F. The Neanderthals and their evolutionary significance. In *The Origins of Modern Humans: A World Survey of the Fossil Evidence*, edited by F. Spencer, 1–49. Alan R. Liss, 1984.

Spencer, F. *Piltdown: A Scientific Forgery*. Oxford University Press, 1990.

Trinkaus, E., and P. Shipman. *The Neanderthals*. Alfred Knopf, 1993.

Chapter 3: Two Models

Aiello, L. C. The fossil evidence for modern human origins: A revised view. *American Anthropologist* 95 (1993):73–96.

Clark, G. A. Continuity or replacement? Putting modern human origins in an evolutionary context. In *The Middle Paleolithic: Adaptation, Behavior and Variability*, edited by H. Dibble and P. Mellars, 183–205. University of Pennsylvania Museum, 1992.

Frayer, D. W., et al. Theories of modern human origins: The paleontological test. *American Anthropologist* 95 (1993):14–50.

Howells, W. W. Explaining modern man: Evolutionists versus migrationists. *Journal of Human Evolution* 5 (1976):477–495.

Mellars, P. Major issues in the emergence of modern humans. *Current Anthropology* 30 (1989):349–385.

Pope, G. G. Recent advances in Far Eastern paleoanthropology. *Annual Review of Anthropology* 17 (1988):43–77.

Smith, F. H. The Neanderthals: Evolutionary dead ends or ancestors of modern people? *Journal of Anthropological Research* 47 (1991):219–238.

Stringer, C. B. The emergence of modern humans. *Scientific American* (December 1990):98–104.

Stringer, C. B., and P. Andrews. Genetic and fossil evidence for the origin of modern humans. *Science* 239 (1988):35–68.

Stringer, C. B., and R. Grün. Time for the last Neanderthals. *Nature* 351 (1991):701–702.

Wolpoff, M. H., et al. Reply to Stringer and Andrews. *Science* 41 (1988):772–773.

Thorne, A., and M. H. Wolpoff. The multiregional evolution of humans. *Scientific American* (April 1992):76–84.

Chapter 4: Mitochondrial Eve

Cavalli-Sforza, L. L. Genes, people and languages. *Scientific American* (November 1991):104–110.

Goldman, N., and N. H. Barton. Genetics and geography. *Nature* 357 (1992):440–441.

Hedges, S. B., et al. Human origins and analysis of mitochondrial DNA sequences. *Science* 255 (1992):737–738.

Johnson, M. J., et al. Radiation of human mitochondrial DNA types analyzed by restriction endonuclease cleavage patterns. *Journal of Molecular Evolution* 19 (1983):255–271.

Maddison, D., et al. Geographic origins of human mitochondrial DNA: Phylogenetic evidence from control region sequences. *Systematic Biology* 41 (1992):111–124.

Merriwether, D. A., et al. The structure of human mitochondrial variation. *Journal of Molecular Evolution* 33 (1991):543–555.

Mountain, J. L., et al. Evolution of modern humans: Evidence from nuclear polymorphisms. *Transactions of the Royal Society B* 337 (1992):159–165.

Stoneking, M., et al. New approaches to dating suggest a recent age for the human mtDNA ancestor. *Transactions of the Royal Society B* 337 (1992): 167–175.

Templeton, A. R., Human origins and analysis of mitochondrial DNA sequences. *Science* 255 (1992):737.

Templeton, A. R. The "Eve" hypothesis: A genetic critique and reanalysis. *American Anthropologist* 95 (1993):51–72.

Thorne, Alan G., and Milford H. Wolpoff. The multiregional evolution of humans. *Scientific American* (April 1992):76–83.

Wilson, A. C., and R. L. Cann. The recent African genesis of humans. *Scientific American* (April 1992):68–73.

Vigilant, L., et al. African populations and the evolution of human mitochondrial DNA. *Science* 253 (1991):1503–1507.

Chapter 5: The Archeology of Modern Humans

Binford, L. Human ancestors: Changing views of their behavior. *Journal of Anthropological Archeology* 4: 292–327.

Clark, G. A., and J. M. Lindly. The case for continuity: Observations on the biocultural transition in Europe and western Asia. In *The Human Revolution*, edited by Paul Mellars and Chris Stringer, 626–676. Princeton University Press, 1989.

Clark, J. D. The origins and spread of modern humans: A broad perspective on the African evidence. In *The Human Revolution*, edited by Paul Mellars and Chris Stringer, 565–588. Princeton University Press, 1989.

Harrold, F. B. Mousterian, Chatelperronian, and early Aurignacian: Continuity or discontinuity? In *The Human Revolution*, edited by Paul Mellars and Chris Stringer, 677–713. Princeton University Press, 1989.

Isaac, G. The archeology of human origins. *Advances in World Archeology* (1984):1–87.

Klein, R. The archeology of modern humans. *Evolutionary Anthropology* 1 (1992):5–14.

Shipman, P. Scavenging or hunting in early hominids. *American Anthropologist* 88 (1986):27–43.

Pope, G. Bamboo and human evolution. *Natural History* (October 1989):49–56.

White, R. Rethinking the Middle/Upper Paleolithic transition. *Current Anthropology* 23 (1982):169–189.

Chapter 6: Symbolism and Images

Bahn, P. Pigments of the imagination. *Nature* 347 (1990):426.

Chase, P. G., and H. L. Dibble. Middle Paleolithic symbolism: A review of current evidence and interpretations. *Journal of Anthropological Archeology* 6 (1987):263–296.

Clark, G. A., and J. M. Lindly. The case for continuity: Observations on the biocultural transition in Europe and western Asia. In *The Human Revolution*, edited by Paul Mellars and Chris Stringer, 626–676. Princeton University Press, 1989.

Clottes, J. Paint analyses from several Magdalenian caves in the Ariège region of France. *Journal of Archeological Science*, in press.

Davidson, I., and W. Noble. The archeology of depiction and language. *Current Anthropology* 30 (1989): 125–156.

D'Errico, F. Technology, motion, and the meaning of epipaleolithic art. *Current Anthropology* 33 (1992): 94–109.

Halverson, J. Art for art's sake in the Paleolithic. *Current Anthropology* 28 (1987):63–89.

Lindly, J. M., and G. A. Clark. Symbolism and modern human origins. *Current Anthropology* 31 (1991): 233–262.

Lewis-Williams, J. D. Cognitive and optical illusions in San rock art research. *Current Anthropology* 27 (1986):171–177.

Lorblanchet, M. Spitting images. *Archeology* (November/December 1991): 27–31.

White, R. Rethinking the Middle/Upper Paleolithic transition. *Current Anthropology* 23 (1982):169–189.

White, R. *Dark Caves, Bright Visions*. The American Museum of Natural History, 1986.

White, R. Visual thinking in the Ice Age. *Scientific American* (July 1989):92–99.

Valladas, H., et al. Direct radiocarbon dates for prehistoric paintings at the Altamira, El Castillo and Niaux caves. *Nature* 357 (1992):68–70.

Chapter 7: Language and Modern Human Origins

Aiello, L. C., and Robin Dunbar. Neocortex size, group size, and the evolution of language in the hominids. In press.

Cheney, D., et al. Social relationships and social cognition. *Science* (1986):1361–1366.

Davidson, I., and W. Noble. The archeology of depiction and language. *Current Anthropology* 30 (1989): 125–156.

Deacon, T. W. The neural circuit underlying primate calls and human language. *Human Evolution* (1989):367–401.

Dunbar, Robin M. Neocortex size as a constraint on group size in primates. *Journal of Human Evolution* 22 (1992):469–493.

Falk, D. 3.5 million years of hominid brain evolution. *Seminars in the Neurosciences* 3 (1991):409–416.

Foley, R. A. Language origins: The silence of the past. *Nature* 353 (1991):114–115.

Frayer, D. W. Language capacity in European Neanderthals. *Current Anthropology*, in press.

Gibson, K., and T. Ingold, eds. *Tools, Language, and Intelligence*. Cambridge University Press, 1992.

Harcourt, A. H. Alliances in contests and social intelligence. In *Social Expertise and the Evolution of Intellect*, edited by R. Byrne and A. Whiten. Oxford University Press, 1988.

Holloway, R. L. Human brain evolution. *Canadian Journal of Anthropology* 3 (1983):215–230.

Holloway, R. L. Human paleontological evidence relevant to language behavior. *Human Neurobiology* 2 (1983):105–114.

Humphrey, N. K. *The Inner Eye.* Faber and Faber, 1986.

Isaac, G. L. Stages of cultural elaboration in the Pleistocene. In *Origins and Evolution of Language and Speech.* New York Academy of Sciences, 1976.

Jerison, H. J. Brain size and the evolution of mind. Fifty-ninth James Arthur Lecture, American Museum of Natural History, 1991.

Laitmann, J. T. The anatomy of human speech. *Natural History* (August 1983):20–27.

Marshack, A. Implications of the Paleolithic evidence for the origin of language. *American Scientist* 64 (1976):136–145.

Martin, R. D. Human brain evolution in an ecological context. Fifty-second James Arthur Lecture, American Museum of Natural History, 1983.

Pagel, M. D., and P. H. Harvey. How mammals produce large-brained offspring. *Evolution* (1988): 948–957.

Pinker, S., and P. Bloom. Natural language and natural selection. *Behavioral and Brain Sciences* 13 (1990):707–784.

Savage-Rumbaugh, S., and D. Rumbaugh. The emergence of language. In *Tools, Language and Cognition*, edited by K. R. Gibson and T. Ingold. Cambridge University Press, 1993.

White, R. Thoughts on social relationships and language in hominid evolution. *Journal of Social and Personal Relationships* 2 (1985):95–115.

Wynn, T. Archeological evidence for modern intelligence. In *The Origin of Human Behavior*, edited by R. A. Foley, 52–66. Unwin Hyman, 1991.

SOURCES OF ILLUSTRATIONS

Page 48
From *Fossil Men* by Marcellin Boule. Oliver and Boyd, 1928, p. 239

Page 50
Natural History Museum, London

Page 52
American Museum of Natural History

Page 57
Royal Geographic Society

Page 60
Peter Kain/Sharma

Page 62
Murray Alcosser

Page 64
Ira Block

Page 66
Collection of Roger Lewin

Page 69
Ira Block

Page 76
Left, Cary Wolinsky/Tony Stone Worldwide
Middle, Catherine Karnow/Woodfin Camp & Assoc.
Right, Hilarie Kavanagh/Tony Stone Worldwide

Page 78
Redrawn from a figure in "The Multiregional Evolution of Humans" by Alan G. Thorne and Milford H. Wolpoff. *Scientific American,* Vol. 266, No.4, April 1992, p. 78

Page 79
Peter Yates

Page 81
Redrawn from a figure in "The Emergence of Modern Humans" by Christopher B. Stringer. *Scientific American,* Vol. 263, No. 6, December 1990, p. 104

Page 82
Adapted from a figure in "The Earliest History of the Earth" by Derek York. *Scientific American,* Vol. 268, No. 1, January 1993, p. 94

Page 84
Redrawn from a figure in "The Emergence of Modern Humans" by Christopher B. Stringer. *Scientific American,* Vol. 263, No. 6, December 1990, p. 100

Page 85
Ira Block

Page 86
Adapted from a map in "The Emergence of Modern Humans" by Christopher B. Stringer. *Scientific American,* Vol. 263, No. 6, December 1990, p. 103

Page 88
Kosi Lum and Rebecca Cann, Dept. of Genetic and Molecular Biology, U. of Hawaii

Page 90
James D. Wilson/Woodfin Camp & Assoc.

Page 91
Douglas Wallace

Page 92
Redrawn from a figure in "The Recent African Genesis of Humans" by Allan C. Wilson and Rebecca L. Cann. *Scientific American,* Vol. 266, No. 4, April 1992, p. 70

Page 96
Adapted from a figure in "The Recent African Genesis of Humans" by Allan C. Wilson and Rebecca L. Cann. *Scientific American,* Vol. 266, No. 4, April 1992, p. 69

Page 101
Redrawn from a graph in "The Recent African Genesis of Humans" by Allan C. Wilson and Rebecca L. Cann. *Scientific American,* Vol. 266, No. 4, April 1992, p. 71

Page 102
Redrawn from a figure in "The Recent African Genesis of Humans" by Allan C. Wilson and Rebecca L. Cann. *Scientific American,* Vol. 266, No. 4, April 1992, p. 72

Page 103
Redrawn from a figure in "The Multiregional Evolution of Humans" by Alan G. Thorne and Milford H. Wolpoff. *Scientific American,* Vol. 266, No. 4, April 1992, p. 83

Page 104
Ed Kashi

Page 106
Redrawn from a figure in "The Emergence of Modern Humans" by Christopher B. Stringer. *Scientific American,* December 1990, p. 102

Page 107
Redrawn from a figure in "Genes, Peoples and Languages" by Luigi

Luca Cavalli-Sforza. *Scientific American,* Vol. 265, No. 5, November 1991, pp. 108-109

Page 110
Redrawn from a figure in "Genes, Peoples and Languages" by Luigi Luca Cavalli-Sforza. *Scientific American,* Vol. 265, No. 5, November 1991

Page 114
Hermitage Museum, St. Petersburg/Sisse Brimberg

Page 118
John Reader/Science Photo Library, Photo Researchers

Page 119
Musée Mas-d'-Azil/Sisse Brimberg

Page 120
Musée des Antiquités Nationales/ Sisse Brimberg

Page 122
Richard Klein

Page 123
Redrawn from a figure in an article by Richard Klein. *Evolutionary Anthropology,* Vol. 1, No. 1, 1992

Page 124
Ira Block

Page 127
H. J. Deacon/Anthony Bannister Photo Library

Page 128
Chip Clark

Page 129
Alison Brooks and John Yellen

Page 130
Ira Block

Page 136
Sisse Brimberg/Norbert Aujoulat

Page 139
John Reader/Science Photo Library, Photo Researchers

Page 141
Randall White

Page 142
Sisse Brimberg

Page 143
Jean-Paul Ferrero/Auscape

Page 144
M. Lariviere. Collection Begouen

Page 145
Sisse Brimberg/Norbert Aujoulat

Page 147
Musée des Antiquités Nationales/ Sisse Brimberg

Page 148
Jean Clottes

Page 149
Sisse Brimberg

Page 152
Rock Art Research Unit, Department of Archaeology, U. of the Withwatersrand, Johannesburg

Page 155
Norbert Aujoulat

Page 157
Alexander Marschack

Page 160
John Reader/Science Photo Library, Photo Researchers

Page 165
Kevin O'Farrell, CONCEPTS

Page 166
Ralph Holloway

Page 170
Courtesy of Jeffrey Laitman

Page 174
Enrico Ferorelli

Page 178
Sovfoto

Page 179
John Reader/Science Photo Library, Photo Researchers

Page 181
Reprinted from *Hundert Jahre Neanderthaler,* G. H. R. Koenigswald, ed., Kemink en Zoon, Utrecht, The Netherlands, 1958, by permission of the Wenner-Gren Foundation for Anthropological Research, Inc., New York

Page 182
Sisse Brimberg

INDEX

Note: Page numbers in italics indicate illustrations.

OTHER BOOKS IN THE SCIENTIFIC AMERICAN LIBRARY SERIES